Low Cost Manufacturing Technologies

Shrikrishna Nandkishor Joshi ·
Uday Shanker Dixit · R. K. Mittal · Swarup Bag
Editors

Low Cost Manufacturing Technologies

Proceedings of NERC 2022

Springer

Editors
Shrikrishna Nandkishor Joshi
Department of Mechanical Engineering
Indian Institute of Technology Guwahati
Guwahati, Assam, India

Uday Shanker Dixit
Department of Mechanical Engineering
Indian Institute of Technology Guwahati
Guwahati, Assam, India

R. K. Mittal
Department of Mechanical Engineering
Indian Institute of Technology Guwahati
Guwahati, Assam, India

Swarup Bag
Department of Mechanical Engineering
Indian Institute of Technology Guwahati
Guwahati, Assam, India

ISBN 978-981-19-8451-8 ISBN 978-981-19-8452-5 (eBook)
https://doi.org/10.1007/978-981-19-8452-5

This Springer imprint is published by the registered company Springer Nature Singapore Pte Ltd.
The registered company address is: 152 Beach Road, #21-01/04 Gateway East, Singapore 189721, Singapore

Foreword

It is a matter of great satisfaction for me that Indian Institute of Technology Guwahati successfully hosted North-East Research Conclave (NERC) 2022 during 20–22 May 2022. The NERC 2022 was conducted on the theme "Sustainable Science and Technology". Concurrently, Assam Biotech Conclave (ABC) was also organized on 21–22 May 2022. Both the events attracted huge participation from policy-makers, researchers, industrialist, army and students. Even the participation of school children was overwhelming.

NERC and ABC had many events including panel discussions, exhibitions, keynote lectures, competitions and paper presentations. Presentation of technical papers forms the core of any research conference. NERC attracted 879 research papers on various themes covering science, technology and humanities. Out of these, some select papers have been published by Springer Nature in the form of 15 volumes. These papers have been peer reviewed and thoroughly edited by IIT Guwahati faculty members. I am sure that these volumes will prove to be excellent resource material for research. Most of the papers presented in these volumes highlight the special needs and aspiration of eight states of North-East India. I congratulate and thank authors, reviewers, editors and publisher for bring out proceedings.

Motivation for organizing NERC came from none other than Honourable Minister of Education, Government of India, Shri Dharmendra Pradhan Ji. It helped to bring policy-makers, researchers, industrialists, academicians, students and children in one forum. It is perhaps the rarest conclave covering almost all possible research themes. For better readability, the proceedings has been divided into 15 volumes, but each volume reflects diversity in terms of topics and researchers. Only common thread is sustainable development of North-East India. Invariably, Sustainable North-East India is a prerequisite for sustainable India and the whole world. In that sense, these 15 volumes will serve guiding and stimulating light for all the stakeholders of the development. I am pleased to dedicate these volumes to nation as a part of Azadi ka Amrit Mahotsav.

T. G. Sitharam
Director
Indian Institute of Technology Guwahati
Guwahati, India

Chairman
All India Council for Technical Education
New Delhi, India

Preface

In 2020–21, the manufacturing sector contributed 14.43% to Gross Value Added in India. Thus, it plays an important role in the industrial growth and development of the nation. Moreover, it creates several employment opportunities and helps India to become a self-reliant (*atmanirbhar*) nation. In line with the efforts of the Government of India to promote sustainable manufacturing, North-East Research Conclave (NERC) 2022 invited research papers on the cutting-edge manufacturing technologies and product development under the track "Low-cost Manufacturing". Low-cost manufacturing works on the principle of optimizing the available resources without compromising the product quality. It comprises the research and development activities that influence the cost of manufacturing of a product or a system. Influencing factors are materials, manufacturing processes, material handling processes, skilled manpower, quality control technologies, effective communication, and use of artificial intelligence techniques.

In this track, 45 extended abstracts were received. Out of them, 39 abstracts were presented in oral-mode and the remaining in poster-mode. These presentations were based on the research carried out on various fields of manufacturing such as computer-aided design and manufacturing (CAD/CAM), additive manufacturing, laser-based manufacturing, novel joining techniques, ultra-precision machining, advanced material forming, nanotechnology, surface coatings and polishing techniques, sensors and instrumentation, and novel product and material development. Both numerical and experimental research findings were shared. Out of the presented papers, 18 high-quality full-length papers have been included in this book volume. We hope that researchers, industry professionals, policy makers, and students will find this collection valuable.

We are thankful to the organizers of NERC (2022), authors, reviewers, and participants. Professor T. G. Sitharam, Director, Indian Institute of Technology Guwahati, has been a source of inspiration. Cooperation provided by Prof. Vimal Katiyar, Convener, and Prof. Subhendu Sekhar Bag, Co-convener, is commendable. We would also like to place on record the excellent support provided by team of Springer Nature that resulted in the timely publication of this volume.

Guwahati, India

Prof. Shrikrishna Nandkishor Joshi
Prof. Uday Shanker Dixit
Dr. R. K. Mittal
Prof. Swarup Bag

About IIT Guwahati

Indian Institute of Technology (IIT) Guwahati established in 1994 has completed 25 years of glorious existence in 2019. At present, the Institute has eleven departments, seven interdisciplinary academic centres and five academic schools covering all the major engineering, science, health care, management and humanities disciplines, offering B.Tech., B.Des., M.A., M.Des., M.Tech., M.Sc. and Ph.D. programmes. The institute presently offers a residential campus to 435 faculty members and more than 7500 students at present. Besides its laurels in teaching and research, IIT Guwahati has been able to fulfil the aspirations of people of the North-East region to a great extent since its inception in 1994. The picturesque campus is on a sprawling 285 hectares plot on the north bank of the Brahmaputra, around 20 km from the heart of the Guwahati city.

IIT Guwahati is the only academic institution in India that occupied a place among the top 100 world universities—under 50 years of age—ranked by the London-based Times Higher Education (THE) in the year 2014 and continues to maintain its superior position even today in various International Rankings. IIT Guwahati gained rank 37 globally in the "Research Citations per Faculty" category and overall 384 rank in the QS World University Rankings 2023 released recently. IIT Guwahati has retained the 7th position among the best engineering institutions of the country in the "India Rankings 2021" declared by the National Institutional Ranking Framework (NIRF) of the Union Ministry of Education. IIT Guwahati has been also ranked 2nd in the "Swachhata Ranking" conducted by the Government of India. Recently, IIT Guwahati has been ranked as the top-ranked University in 2019 for IT developers by HackerRank in the Asia-Pacific region.

Among other frontier areas of research and innovation, IIT Guwahati is working towards augmenting critical science research initiatives in genomics, developmental biology, health care and bioinformatics, flexible electronics, advanced functional materials, sustainable polymers, rural technologies, renewable energy, artificial intelligence, disaster resilience and risk reduction and water resources and management. In its silver jubilee year, IIT Guwahati is poised to scale newer heights through all-round growth and development.

Indian Institute of Technology Guwahati has dedicated itself to the cause of improving and empowering Northeast India through cutting-edge research, region relevant projects, innovations, individual and multilateral collaborations and special initiatives. Being the only IIT in the entire Northeastern region, IIT Guwahati has an immense amount of responsibility to develop the region and empower the people of the region.

While the entire country is celebrating the "Azadi ka Amrit Mahotsav"—75 glorious years of Independence, and the great pride with which our nation of more than a billion people has been steadily growing today, IIT Guwahati is strongly committed to support that pace of growth for the entire NE so that we can keep pace along with the rest of the country. The specific areas of focus where IIT Guwahati has been contributing immensely to the region are:

(a) Infrastructure development across multiple sectors
(b) Providing solutions for multiple natural disasters such as recurring floods, landslides, earthquakes, cyclones, hailstorms and other natural calamities
(c) Improving the education sector and creating opportunities for employment
(d) Internet, telecommunication and cultural integration
(e) Technological intervention in interdisciplinary areas
(f) Healthcare services and education
(g) Renewable energy generation (solar, wind, biomass, hydro, geothermal)
(h) Overall industrialization, refining fossil fuels and setting up biorefineries.

Besides bringing in the state-of-the-art technical knowhow for most of the above sectors, the institute has been partnering with the local governments and enhancing the technological and educational interactions such that the next-generation youth are empowered with knowledge, skills and necessary entrepreneurial ability. These measures in Assam as well as all other northeast states will usher in a new era of growth, and the opportunities it will provide for interaction with the ASEAN countries as part of the Act East Policy of the Government of India will bring prosperity to this region.

Prof. Parameswar K. Iyer
Dean, Public Relations, Branding and Ranking
Indian Institute of Technology Guwahati

From the Desk of Chairman of Technical Committee of NERC 2022

North-East Research Conclave 2022 was successfully organized during 20–22 May 2022 with the participation of thousands of delegates. A total of 879 oral and poster papers were presented in the conference on 16 different tracks. The theme of the conclave was Sustainable Science and Technology, which is very pertinent in the modern era of globalization. Science and technology has to address economic, environmental and social problems of the world. Technology and sustainability are not incompatible. In fact, technology can achieve the goal of sustainability, which also includes preserving our rich cultural heritage. Concurrently with North-East Research Conclave (NERC), Assam Biotech Conclave 2022 was also organized on 21–22 May 2022. These mega events were organized at Indian Institute of Technology Guwahati (IITG) in physical mode after two years of pandemic period. Along with IITG, Science, Technology and Climate Change Department and Department of Education, Government of Assam were also organizers of these events under the patronage of Shri Dharmendra Pradhan Ji, Honourable Minister of Education and Minister of Skill Development and Entrepreneurship in the Government of India, and Shri Himanta Biswa Sarma Ji, Honourable Chief Minister of Assam.

It is a matter of great pleasure that Springer Nature is publishing the select papers from the conclave in 15 volumes. These are Advanced Functional Materials, Low Cost Manufacturing Technologies, Agro and Food Processing Technologies, Artificial Intelligence and Data Science based R&D interventions, Conservation of Biodiversity in the North Eastern States of India, Disaster Management, Healthcare Research and Related Technologies, Innovative Design for Societal Needs, Policies for Research and Innovation, Research and Innovation for Sustainable Development Goals, Sustainable Environment, Sustainable Energy Generation and Storage, Sustainable Transportation and Urban Development, Teaching and Learning Technologies, Technologies for Rural Development. These volumes are useful archival and reference materials for policy-makers, researchers and students.

As Chairman of Technical Committee, I am thankful to all Editors of all volumes, reviewers and student volunteers who have put tireless efforts to review, select and edit the papers of respective divisions, overcoming the time-constraint. Support provided by Convener, Prof. Vimal Katiyar, Dean R&D, IITG, and Co-conveners

Prof. Subhendu Sekhar Bag, Associate Dean R&D, IITG, and Shri Kailash Karthik N., IAS, is commendable. It is difficult to express words of gratitude for the Director, IITG, Prof. T. G. Sitharam, who has been motivating and guiding all the teams of NERC 2022 and ABC 2022.

Uday Shanker Dixit
Professor, Department of Mechanical Engineering, and
Head, Center for Indian Knowledge Systems
Indian Institute of Technology Guwahati

North East Research Conclave-2022: Toward Sustainable Science and Technology

It is extremely important and imperative to have knowledge-driven growth based on innovation in the case of academic higher education institutes of high repute. The North-Eastern region endowed with rich biodiversity comprises eight states. However, the climatic conditions, limited connectivity, lack of research infrastructure/institutes, territorial conflicts and the mountainous terrain of these regions are major impediments to the research ecosystem in the North-East. Quality higher education focusing on industry–academia collaboration and translational research is extremely beneficial for society. It has also been rightly pointed out by the Hon'ble Prime Minister Sh. Narendra Modi that, "*India cannot develop till Eastern India develops*".

With this idea and as India marks 75 years of Independence, Indian Institute of Technology Guwahati organized "The North-Eastern Research Conclave" from 20 to 22 May 2022. This grand event was jointly conducted with Science, Technology and Climate Change Department and the Department of Education, Government of Assam, at IIT Guwahati Campus.

The mission behind the conclave was to showcase the best R&D activities from educational and research institutions across North-East India and to create an environment, conducive to development of local indigenous technologies and innovations, creating the scope and laying the foundation for entrepreneurship.

In order to attract people and spread awareness about the event, a roadshow was initiated from IIT Guwahati on 7 May 2022 in order to reach all the partnering academic institutes and make them an integral part of the mega event. The Director, IITG, waved the NERC 2022 flag and sent off the road show vehicle from the institute. More than 400 students, staff and faculty participated actively in the roadshow.

A huge response was received by participants from throughout the country. The total no. of participating institutions in this conclave included 7 IITs, 10 NITs, 5 IIITs and other CFTIs, 23 research laboratories, 17 central-funded universities, 47 other universities/institutes along with about 100 schools. Eminent personalities from industries, start-ups, research councils and PSUs also joined in.

The presence of dignitaries from important Ministries was observed such as Shri Dharmendra Pradhan, Hon'ble Union Minister of Education and Minister of Skill Development and Entrepreneurship, Government of India; Dr. Himanta Biswa

Sarma, Hon'ble Chief Minister of Assam State; Dr. Ranoj Pegu, Hon'ble Minister of Education, Government of Assam; Dr. Rajkumar Ranjan Singh, Hon'ble Minister of State for Education, Government of India; Dr. Subhas Sarkar, Hon'ble Minister of State for Education, Government of India; Shri Keshab Mahanta, Hon'ble Minister of Science Technology and Climate Change, Government of Assam and many more.

The inauguration ceremony of the conclave was followed by the signing of an MoU between IIT Guwahati and the Government of Assam to establish "The Assam Advanced Health Innovation Institute (AAHII)". This MoU would prove to be a unique partnership between the Government of Assam and IIT Guwahati in order to set up a research institution to leverage advanced technologies to transform medical science. This joint venture company will be able to invite participation from intending parties including corporates/businesses/research institutions and philanthropic organizations.

The third edition of Assam Biotech Conclave 2022 was also held as part of NERC 2022. It brought together the Biotech Entrepreneurs, industry leaders, researchers, academicians, government representatives, policy-makers, innovators and investors together on one platform to explore the possibilities of biotechnology in North-East India and to discuss the new opportunities in the transition.

Officers from the Indian Army also actively participated in the conclave. A talk on "Atmanirbhar Bharat—Indian Army Initiatives towards Self Reliance" was delivered by Lt. Gen. D. S. Rana AVSM, YSM, SM General Officer Commanding, Gajraj Corps on 21 May 2022. The talk was aligned with the vision of the apex leadership of the Government of India and initiatives undertaken by the Indian Armed Forces with a focus on the integration of civil–military establishment in the field of self-reliance. He also elucidated that institutions such as IIT Guwahati which has many running research projects and elaborate student exchange and joint collaboration setup with a large number of countries have the wherewithal to take up defence-related R&D and also facilitate delivery with industry partners. He also invited IIT Guwahati to participate in EAST TECH Symposium planned at Kolkata in July 2022. This led to the signing of an MoU between Indian Army Eastern Command and IIT Guwahati on 7 July 2022 during East Tech 2022. This would further impetus to Indigenisation and Raksha Atmanirbharta.

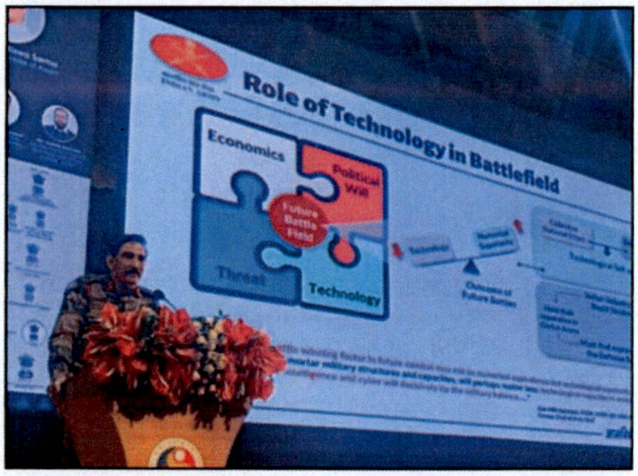

Royal Society of Chemistry, Global battery experiment was performed by more than 1300 students in three sessions starting from 20 May to 22 May at IIT Guwahati. Along with the global battery experiment, creating skilful educators (teacher training programme) was also conducted in parallel sessions. Students had arrived from various schools across Assam and other North-Eastern states.

The Guwahati Declaration was launched at the valedictory ceremony of the conclave by Shri Lok Ranjan, Secretary, Ministry of Development of North Eastern Region (DoNER), in the presence of Shri Kailash Karthik, Deputy Commissioner, Kamrup. The declaration is intended to create a set of guidelines, through which individual as well as a collective responsibility to promote and encourage innovation at the grass-root level and strive to stimulate and execute indigenization and entrepreneurship can be taken up.

Science, education, research and innovation are the four pillars on which the development, as well as the work culture of a nation, rests. This was well articulated by the promising number of exhibitors being seen participating from all across the NE states in the NERC 2022. All the NITs, CFTIs and CFIs were allocated two stalls each, where the delegates showcased the working models of their inventions. Distinctive pavilions were arranged for IIT, NIT, CFIs and CFTIs. Excellent response was obtained from the start-ups all across the NE states. Federation of Industry Commerce of North Eastern Region (FINER) had partnered with NERC 2022 as an industry partner, and they showcased 50 start-ups as a part of the exhibition under the FINER Pavilion. Other significant organizations that came forward to showcase their allied R&D start-ups were the Oil and Natural Gas (Oil and Natural Gas Pavilion), Indian Army (Defense Pavilion) and NE-Railway (NE-Railway Pavilion).

Multifarious research work on topics of societal relevance was presented by researchers from different organizations/institutes. The presentations were conducted in oral and poster presentation modes. The thematic areas for these presentations were part of some of the Sustainable Development Goals (SDGs) such as SDG-3: Good Health and Wellbeing; SDG-7: Affordable and Clean Energy; SDG-9: Industry, Innovation and Infrastructure; SDG-11: Sustainable Cities and Communities and SDG-12: Responsible Consumption and Production. Some of the papers highlighted environmental sustainability, efficiency and management issues, which are important to be presented in the case of North-East regions. Two awards were given under each technical category for these presentations. Overall, the technical sessions were a grand success due to the active cooperation from editors, chairpersons of all the sessions and student volunteers of IITG.

The Government of India has taken various steps to encourage women in the field of science and technology. In this line, the IIT Guwahati Woman Researcher Award was approved to recognize the contribution of women Faculty members of IIT Guwahati fraternity. This prestigious award was conferred to Dr. Latha Rangan who is Senior Professor in the Department of Biosciences and Bioengineering, Indian Institute of Technology Guwahati, India. Prof. Rangan has played a key role in plant biotechnology and sustainable development and especially in the areas of energy security, food security and medicinal crops.

The conclave paved the way for creating mass awareness of Research and Innovation for developing a sustainable society. There was knowledge exchange and dissemination that led to the establishment of Centres of Excellence in Translational Collaborative Research and Innovation. This mega event led to the bridging of the gap between industry–academia and creating handholding pathways for setting up long-term collaboration for R&D innovations towards the goal of establishing sustainable NE India. The conclave brought together over 8000 participants including Hon'ble Ministers, Official Bureaucrats, Eminent Professors, Scientists, Renowned Industrialist, School Children/Teachers and Others delegates. This revolutionized the R&D road map of all the NE states through various dissemination of policies which will benefit the sustainable development of all NE states in near future.

It is an honour and a moment of extreme pride for getting the NERC proceedings published in the prestigious Springer volumes. We would like to thank and acknowledge the globally active publisher Springer for helping us being able to publish the articles on 15 broad areas. We would also like to thank all the authors for their contribution to the grand success of NERC 2022 and wish them great success in all of their future endeavours.

Prof. Vimal Katiyar
Dean, R&D
Department of Chemical Engineering
Centre for the Sustainable polymer
Indian Institute of Technology Guwahati
Guwahati, India
vkatiyar@iitg.ac.in

Prof. Subhendu Sekhar Bag
Associate Dean, R&D
Department of Chemistry
Centre for the Environment
Indian Institute of Technology Guwahati
Guwahati, India
ssbag75@iitg.ac.in

Contents

About the Editors

Dr. Shrikrishna Nandkishor Joshi is working as a Professor of Mechanical Engineering at Indian Institute of Technology Guwahati, India. He was a Visiting Faculty at the Asian Institute of Technology, Bangkok, Thailand in 2015. His research interests include advanced and precision manufacturing processes with focus on applications of lasers in manufacturing, thin-wall machining, single point diamond turning, CAD/CAM and manufacturing automation. He has published more than 70 research papers and sixteen book chapters in various refereed International Journals and Conferences and has co-edited three books on various topics such as Advances in Computational Methods in Manufacturing (Springer Singapore 2019), Application of Lasers in Manufacturing (Springer Singapore 2019), and Laser based manufacturing (Springer New Delhi 2015). Seven Ph.D. students have completed their doctoral studies under his supervision; and nine Ph.D. projects are on-going. Prof. Joshi has conducted sponsored research and consultancy projects worth more than Rs. One crore. He has developed a SWAYAM MOOC course on "Automation in Manufacturing" and a web course on "Mechatronics and Manufacturing Automation" for National Programme on Technology Enhanced Learning (NPTEL) funded by Ministry of Human Resource Development, Government of India. He is an Associate Editor of International Journal of Additive and Subtractive Mnaufacturing, Inderscience.

Dr. Uday Shanker Dixit is presently Professor in the department of Mechanical Engineering and Head of Center for Indian Knowledge System at Indian Institute of Technology Guwahati. He received B.E. degree in Mechanical Engineering from erstwhile University of Roorkee (now Indian Institute of Technology Roorkee) in 1987, M.Tech. degree in Mechanical Engineering from Indian Institute of Technology Kanpur in 1993, and Ph.D. in Mechanical Engineering from IIT Kanpur in 1998. He has worked in two industries—HMT, Pinjore and INDOMAG Steel Technology, New Delhi, where his main responsibility was designing various machines. Dr. Dixit joined the Department of Mechanical Engineering, Indian Institute of Technology Guwahati, in 1998, where he is currently a Professor. He was also the Officiating Director of Central Institute of Technology, Kokrajhar from February 2014 to May 2015. Dr. Dixit is actively engaged in research in various areas of design and manufacturing since last three decades. Most of his research work is focused on Modelling of Manufacturing Processes and involves finite element analysis of elasto-plastic problems. He has also contributed in design optimization. He has extensively used finite element and soft computing tools for the design and manufacture of mechatronic systems. He has authored/co-authored 140 journal papers, 139 conference papers, 42 book chapters and 7 books in mechanical engineering. He has also co-edited 9 books related to manufacturing. He has guest-edited 11 special issues of journals. Presently he is an Associate Editor of the *Journal of Institution of Engineers* (India), Series C and Regional Editor Asia of *International Journal of Mechatronics and Manufacturing Systems*. He has guided 16 doctoral and 54 masters' students. Dr. Dixit has investigated a number of sponsored projects and developed several courses. He is a Fellow of Indian Welding Society, National Advisory Committee member of International and All India Manufacturing, Technology and Design Conference. He was a Board Member of Indian Institute of Technology Kanpur during 2018–21. He is reviewer of several journals and has organized several conferences. He has international collaboration with several researchers.

Dr. R. K. Mittal is an Assistant Professor in the Department of Mechanical Engineering at Indian Institute of Technology Guwahati, India. His key research areas are dynamics of machining process, experimental characterization, and modeling of chatter phenomenon in machining, precision machine design and development, and multiscale manufacturing. He has also worked as an Inspire Faculty at Indian Institute of Science Bangalore. He has four years of experience working as a Design Engineer in Bharat Heavy Electricals Ltd., India. He has collaborated with different industries and research institutes for research work and technology development.

Dr. Swarup Bag completed his Bachelors' of Engineering in Mechanical Engineering from the Jalpaiguri Government Engineering College, Jalpaiguri, West Bengal, India in the year 2000, M.E. in Mechanical Engineering in 2002 from Bengal Engineering College, Shibpur, India (now Indian Institute of Engineering Science and Technology) in the year 2002, and Ph.D. on "Development of Bi-directional Heat Transfer and Fluid Flow Model for Reliable Design of GTA and Laser Welding Processes" in 2009 from IIT Bombay. Later he has worked at the Center for Material Forming (CEMEF), MINES Paris Tech, France for 20 months in Metallurgy, Structure and Rheology (MSR) group. In the year 2011, he joined the Department of Mechanical Engineering, Indian Institute of Technology Guwahati as a faculty member. He worked as a visiting scholar at Center for Advanced Materials Joining, University of Waterloo, Canada from May to July 2018. His current area of research is fundamental process modeling of welding and joining technologies, additive manufacturing process and laser material processing. Dr. Bag has published about 69 journal papers, 50 conference papers, and 21 book chapters related to welding and joining processes. He is the author of the book *Computational Models for GTA and Laser Welding Processes* and recipient of 'Royal Arc Award 2009' from Indian Institute of Welding for the best Ph.D. thesis in welding.

Dr. Bag has developed four MOOCs-NPTEL courses and taught nine different theory courses to undergraduate and postgraduate students at his working place. Until now, seven students have received doctorate degree and seven research scholars are working under his guidance. Twenty-two M.Tech. projects and 11 B.Tech. projects have completed so far. Three research projects have completed so far in his credit and two sponsored projects are going on.

Analysis of Numerical Method for Modal Analysis of Thin-Walled Structures for Achieving Low-Cost Manufacturing

S. Gururaja, Kundan K. Singh, and R. K. Mittal

1 Introduction

Thin-walled structures like microelectrodes, microfins, micro-needles, pins, micro-nozzles, etc., are extensively used in various sectors like automobile, avionics, energy, bio-medical and electronics. These high aspect ratio complex 3D thin structures can be fabricated by micromilling process [1, 2]. But, in micromilling, stiffness of the cutting tool will be approximately equal to the stiffness of the thin-walled workpiece to be manufactured. This low stiffness of cutting tool combined with low stiffness of workpiece makes the micromilling process unstable and leads to chatter. Additionally, due to low rigidity of the thin-walled workpiece, the radial cutting forces change instantaneously, which causes the workpiece to deform elastically in addition to the deformation of cutting tool. Thus, the stiffness of the thin-walled workpiece varies after each instance of material removal. Machining of thin-walled structures induces notable changes in the mass and stiffness of the structure while machining. Each cutting operation increases the flexibility of the workpiece while reducing the net mass in terms of material removal in form of chips. The changes in the static and dynamic properties of thin-walled workpiece due to the cutting operation have been conventionally estimated at particular point where cutting tool and workpiece interacts, along the cutting direction either by using the experimental modal analysis (EMA) or by using finite element (FE) methods. EMA can be used to predict the vibration behaviour of structures at specific points along the tool path [3] and also to predict the surface error and chatter stability in milling of thin rectangular plate

S. Gururaja · K. K. Singh (✉)
Department of Mechanical Engineering, BITS-Pilani, Hyderabad Campus, Telangana, India
e-mail: ksingh@hyderabad.bits-pilani.ac.in

R. K. Mittal
Department of Mechanical Engineering, Indian Institute of Technology Guwahati, Guwahati, Assam, India
e-mail: rkmittal@iitg.ac.in

S. N. Joshi et al. (eds.), *Low Cost Manufacturing Technologies*,
https://doi.org/10.1007/978-981-19-8452-5_1

structure [4]. However, the variation of the modal parameters with time (as machining progress) has not been taken into account in both the cases [3, 4]. Estimation of frequency response functions (FRFs) of in-process thin-walled structures is tedious and not feasible for industrial applications such as closely spaced integral blades of rotor (IBR). The change in compliance, while turning thin-walled cylindrical [5–8] and slender parts [9, 10], was studied by modelling the workpiece at different thickness. Lorong et al. [7] assumed that the mode frequencies do not change during cutting operation. Hence, a large deviation was identified between the experimental and the estimated chatter frequencies. Fischer et al. [5] showed experimentally that the flexible structures undergo change in stability lobes due to varying dynamics of the thin-walled workpiece, and also, Lorong et al. [7] proposed spindle speed scheduling for turning the flexible thin-walled disc to reduce chatter. To study the influence of material removal on the dynamic stiffness of the thin-walled structures, the changes in the deflections of the thin-walled structures were predicted during peripheral milling. The predicted deflection was a function of the cutting force during peripheral milling [11–16]. A change in workpiece geometry was introduced by changing the model geometry or re-meshing the FE model [11–15] or by changing the stiffness matrices [16]. Koike et al. [17] used cantilever beam geometry to find the direction of high stiffness by modelling the initial and final workpiece geometry. This helped in reducing the static deflections by changing the direction of cutting forces along the direction of high stiffness. Out of indefinite modes that are excited while machining thin wall, only dominant modes [18–20] in the region of cutting were taken into consideration for predicting stability. In few studies [21, 22] the change in natural frequency was estimated by interpolating the dynamic properties of the intermediate workpiece states along the tool path [23–25]. Few researchers assumed that the vibration characteristics will not change as the material removal in finishing operation is less. The individual stability lobe diagram was generated at specific points along the tool path and were superimposed to get a unified stability lobe [25]. This practice of superimposing is not appreciable as it is practically impossible to generate stability lobes at all discrete points along the tool path. Kersting and Biermann [26] compared three different numerical methods to incorporate the varying thin-walled dynamics to predict deflections in the milling of thin-walled aerospace parts.

From all the literatures, it is very evident that the varying stiffness of the thin-walled workpiece should be considered while predicting stable process parameters. Experimental modal analysis can be used to find the modal parameters of the thin-walled structure. But, the technique cannot be performed for all possible machining conditions as it increases the estimation time and cost of the modal testing. Numerical methods can be a viable replacement for the experimental modal analysis, as experimental modal analysis cannot be performed for all machining conditions. The three major issues of the micromilling are chatter, deflection and surface form errors. All these errors are linked to modal parameters of the workpiece and the cutting tool. Specially in case of thin-walled structures, stiffness of the workpiece changes instantly, which alters the dynamics of the micromilling process. The chosen stable process parameters at the beginning of cutting operation may become unstable process parameter as cutting progress. Hence, the present work proposes numerical simulation to generate stability lobe diagram by incorporating the effect of varying stiffness of the thin-walled structures. It is expected that the process parameters taken from the proposed stability lobes will increase the tool life and improve the surface quality of the micromilled part.

2 Methodology

The cutting coefficients required for the present work are obtained by linear fitment of root mean square (RMS) values of obtained cutting forces. These cutting coefficients are used for predicting stability in high-speed micromilling of flexible thin-walled Ti6Al4V by incorporating it in the two-degree of freedom (DoF) chatter model. The modal parameters are obtained from the simulated frequency analysis in commercial finite element software package after the mesh convergence test for the first mode. The obtained modal parameters for different machining conditions are used to generate the stability lobes, incorporating the effect of varying stiffness of flexible thin-walled structure. The developed stability lobes have been validated by conducting experimental modal analysis on thin-walled Ti6Al4V at different machining conditions. The stability lobes for different machining conditions has been generated by solving the Eigen value problem. The adopted methodology is represented as a block diagram which is given in Fig. 1.

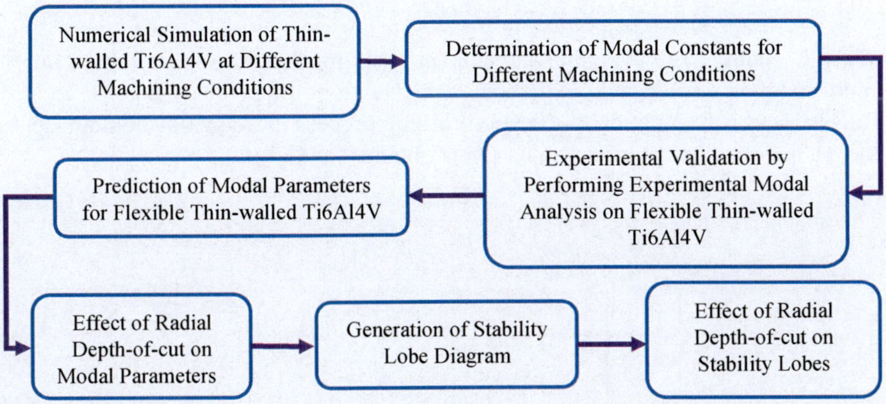

Fig. 1 Implemented methodology for different machining conditions

3 Chatter Modelling

The flexible thin-walled workpiece has been modelled as a 2-DoF spring–mass–dashpot system as given by Fig. 2a. The micro-end mill (Fig. 2b) excites the workpiece by generating the cutting forces that results in deflection of workpiece. A thin-walled workpiece under the influence of the cutting forces can be modelled as a flexible body, and the equations of motion are given as:

$$\ddot{x} + 2\zeta_x\omega_{nx}\dot{x} + \omega_{nx}^2 x = \frac{\omega_{nx}^2}{k_x}\sum_{j=1}^{N} F_{x,j},\ \ddot{y} + 2\zeta_y\omega_{ny}\dot{y} + \omega_{ny}^2 y = \frac{\omega_{ny}^2}{k_y}\sum_{j=1}^{N} F_{y,j} \tag{1}$$

where k_x, k_y, ω_{nx}, ω_{ny}, ζ_x, ζ_y and $F_{x,j}$, $F_{y,j}$ represent the stiffness, natural frequency, damping factor and cutting forces in two normal directions (x and y) on tooth j, respectively. The total number of cutting flutes in the micro-end mill is represented by N, and j represents the instantaneous cutting tooth. F_r and F_t are the total radial and tangential cutting forces on the given tool for all N flutes and are given as:

$$F_r = \sum_{j=1}^{N} F_{r,j}\ F_t = \sum_{j=1}^{N} F_{t,j} \tag{2}$$

where $F_{r,j}$ and $F_{t,j}$ are the radial and tangential cutting forces, respectively, at the jth tooth, and their corresponding mathematical equations are given by Eqs. (4) and (5). The characteristic equation of the micromilling process in frequency domain can be solved, and the characteristics equation is obtained as [27]:

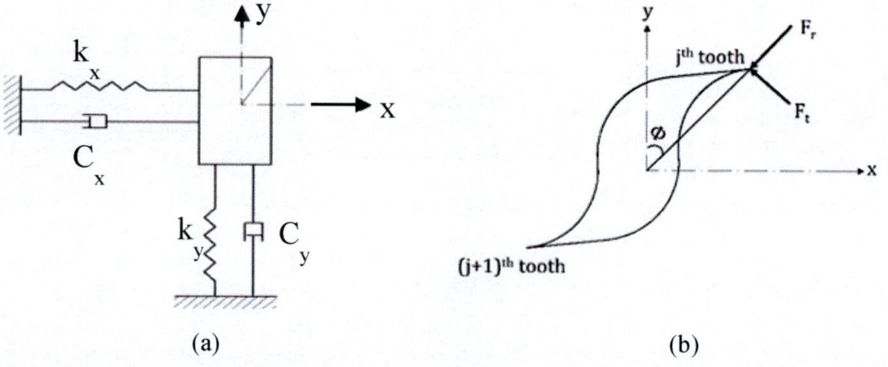

Fig. 2 Schematic representation; **a** thin-walled workpiece as 2-DOF model and **b** model of micro-end mill showing cutting forces at the tool-tip

$$\det\left|[I] - \frac{1}{2}K_{tc}a\left(1 - e^{-j\omega_c T}\right)[A_o][\Phi(i\omega_c)]\right| = 0 \tag{3}$$

where I represents the identity matrix and a represents the stable depth of cut, ω_c is the frequency of chatter, T is the time delay for sequential cuts, A_o is directional cutting coefficient matrix, and $\phi(i\omega_c)$ is the frequency response function (FRF) of thin-walled workpiece.

4 Determination of Cutting Coefficients

Cutting coefficients are obtained by plotting root mean square (RMS) values of radial and tangential cutting forces as a function of uncut chip thickness by performing slotting experiments at different machining conditions. Slotting experiments were done on Ti6Al4V to obtain the cutting coefficients. Micro-end mill used was two-fluted uncoated tungsten carbide of 500 μm diameter and 38 mm overall length. For all instances, 20 mm micro-end mill overhang was maintained and all the cutting operations were performed in dry condition only. Spindle speeds for experimental conditions were changed at intervals of 10,000 rpm from 10,000 to 80,000 rpm at different feed rates. A linear fitment was made between the radial and tangential cutting forces per unit depth of cut and average chip thickness. The radial and tangential cutting forces are given by:

$$F_{r,j} = K_{rc}a\overline{h} \tag{4}$$

$$F_{t,j} = K_{tc}a\overline{h} \tag{5}$$

$\overline{h}$ is the mean chip thickness, which depends on the engagement angle and cutting conditions. In slot milling, mean chip thickness $\overline{h}$ for given feed per tooth (f_t) is taken to be:

$$\overline{h} = \frac{2}{\pi}f_t \tag{6}$$

The linear fitment is obtained by minimizing the least square error, and it is given by:

$$e = \sum_{q=1}^{n}\sum_{p=1}^{m}\left(F_{\text{exp}} - F_{\text{theo}}\right)^2 \tag{7}$$

where n represents the number of iterations for each feed, and m represents the quantity of samples taken for the evaluation. F_{exp} and F_{theo} are the RMS values of experimental and theoretical cutting forces, respectively. The variation of linearly

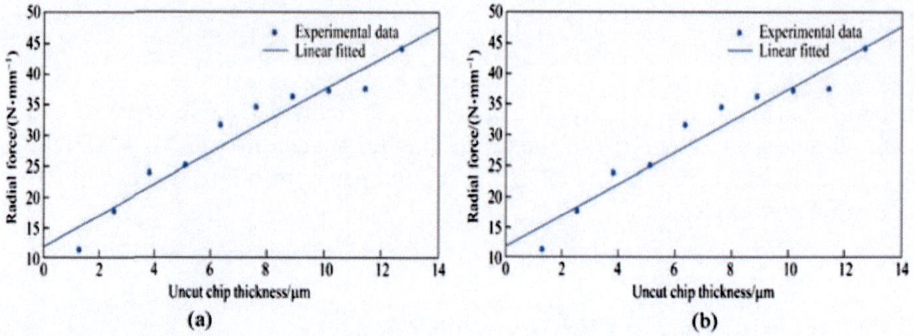

Fig. 3 RMS value of cutting forces against different uncut chip thickness for 80,000 rpm; **a** radial forces versus uncut chip thickness, **b** tangential forces versus uncut chip thickness

Table 1 Cutting coefficients for flexible thin-walled structure

Thin-walled material	Tangential cutting coefficient, K_{tc}, $N \cdot mm^{-2}$	Radial cutting coefficient, K_{tc}, $N \cdot mm^{-2}$
Ti6Al4V	2800	2500

fitted radial and tangential cutting forces with average chip thickness is shown in Fig. 3a, b, respectively. The values of the cutting coefficients are given in Table 1.

5 Estimation of Modal Parameters for Thin-Walled Workpiece

5.1 *Experimental Set-Up*

Experiments were performed at BITS-Pilani, Hyderabad Campus. The 3-axis micro-machining centre with x and y linear stages is driven by AC servomotor with linear encoder. Linear stages (x and y) have a resolution of 0.15 μm and accuracy of ± 2 μm. The maximum spindle speed is 80,000 rpm, and the high-speed spindle is cooled by oil-air mixture, and it has a maximum torque of 11.9 N-cm. The Z-stage is also driven by a AC servomotor with linear encoder, and it houses the high-speed spindle. All three stages are controlled by a tri-axial controller with RS232/USB interface. The high-speed machining centre is mounted on a specially designed granite structure to damp any vibrations in the system while machining. The schematic of the set-up and image of the high-speed machining centre are shown in Fig. 4a, b, respectively.

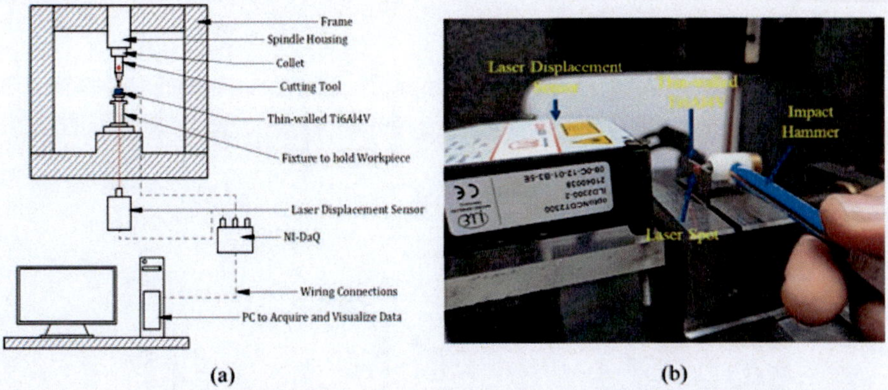

Fig. 4 Set-up for performing experimental modal analysis; **a** schematic of set-up, **b** modal analysis set-up

5.2 *Numerical Modelling*

Estimation of modal parameters is done by performing frequency analysis of flexible thin-walled workpiece using finite element method. The thin-walled workpiece has been modelled as an Euler–Bernoulli beam. An Eigen value analysis has been carried out for the global matrices (mass matrix and stiffness matrix) obtained from the finite element modelling of the Euler–Bernoulli beam of thin-walled workpiece. Estimated Eigen value and Eigen vector have been used to decouple the coupled equation for all degrees of freedom in order to obtain the modal parameters of the thin-walled workpiece. ABAQUS® has been used to verify the different modes frequency by modelling the thin-walled structure as a three-dimensional solid structure. Mesh independence study has been done by comparing the Mode-1 natural frequency in ABAQUS® by varying the mesh size. The optimum mesh size was found to be 0.001 beyond which very minimal change in natural frequency was observed. Figure 5 shows the 3D CAD model of flexible thin-walled Ti6Al4V under different machining conditions. For all the machining cases, a constant axial depth of cut of 50 μm and varying radial depth of cut in steps of 50 μm was varied from 40 to 140 μm as shown in Table 2. The algorithm for the numerical simulation is given in Fig. 6. The flexible thin-walled structure was treated as a cantilever beam, where the bottom portion was fixed rigidly and the modal parameters like natural frequency and stiffness was estimated at the free end. The pre-processing parameters used in numerical simulation are given in Table 3.

The global matrices (mass matrix and stiffness matrix) are obtained from the finite element modelling of the thin-walled workpiece as Euler–Bernoulli beam. The Eigen analysis has been carried out on the global mass and stiffness matrix to get the natural frequency and the modes vector. These modes vectors have been used to obtain the modal constant. These modal constants are then used to find the frequency response function of the thin-walled workpiece at free end. Using the frequency response

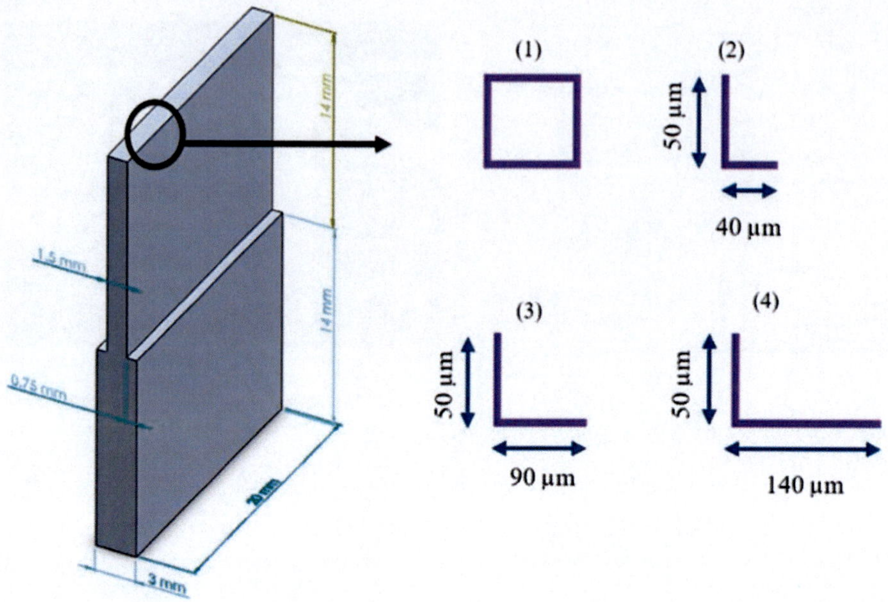

Fig. 5 Thin-walled flexible Ti6Al4V workpiece with different machining conditions

Table 2 Machining condition for numerical simulation and experimentation

Machining condition	Radial depth of cut in µm	Axial depth of cut in µm	Workpiece condition
1	0	50	Static
2	40		
3	90		
4	140		

function (FRFs), the modal stiffness is extracted through peak-picking method. The pseudocode for the estimation of modal stiffness and modal frequency is given in Fig. 7. Since the damping in the flexible thin-walled structure cannot be modelled like stiffness and mass, a proportional damping model is used for the estimation of modal parameters as the modes shape of system with proportional damping is equivalent to the mode shape of undamped system. As the analysis is the case of free vibration, the governing equation for motion of the system with n degree of freedom is:

$$[M]\{\ddot{x}\} + [C]\{\dot{x}\} + [K]\{x\} = \{0\} \tag{8}$$

where $[M]$ and $[K]$ are global mass and stiffness matrices and $[C]$ is the damping matrix of positive definite value to uncouple the Eq. (8). The global matrices (mass

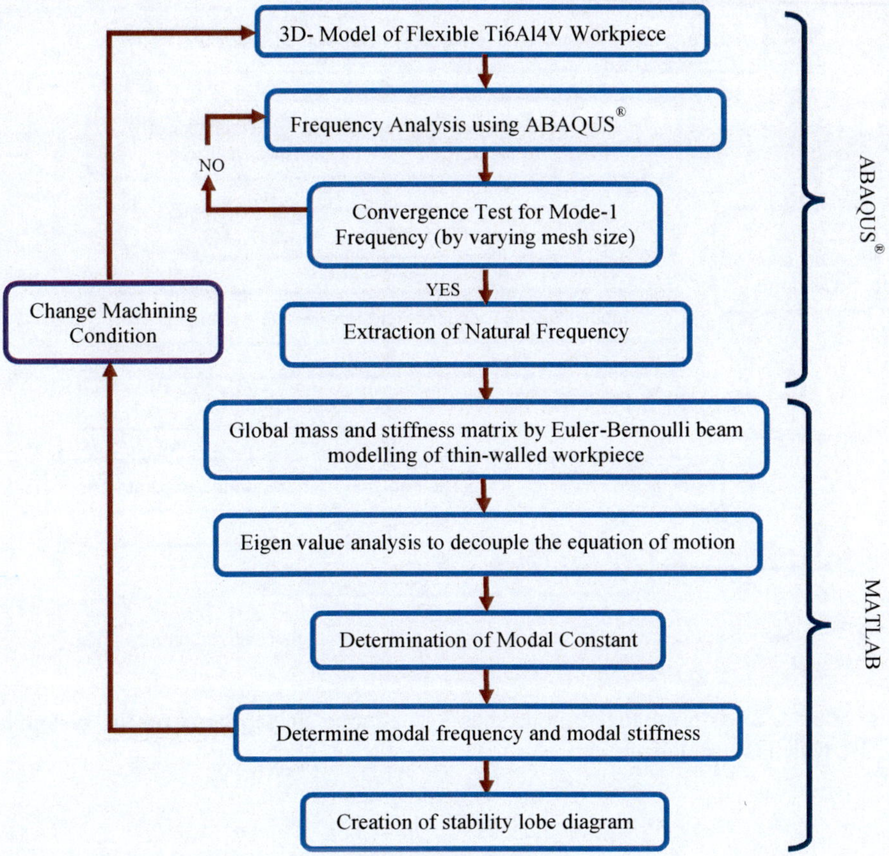

Fig. 6 Algorithm for numerical simulation

Table 3 Pre-processing data of model from ABAQUS

Constant axial depth of cut—50 μm				
Mesh with C3D8 brick element				
Mesh size-0.001	Radial depth of cut 0 μm	Radial depth of cut 40 μm	Radial depth of cut 90 μm	Radial depth of cut 140 μm
No. of elements	9027	9416	9282	9198
Elastic modulus	110 GPa			
Mass density	4470.5 kg/m^3			

matrix and stiffness matrix) are multiplied by mode shape matrix [Ψ], which results in diagonalization of these matrices. Also, as there exists multiple mode shapes for a given system which are all valid, mass-normalized mode shapes are very helpful, as these are unique to the given system. Mode shape normalized using the modal mass

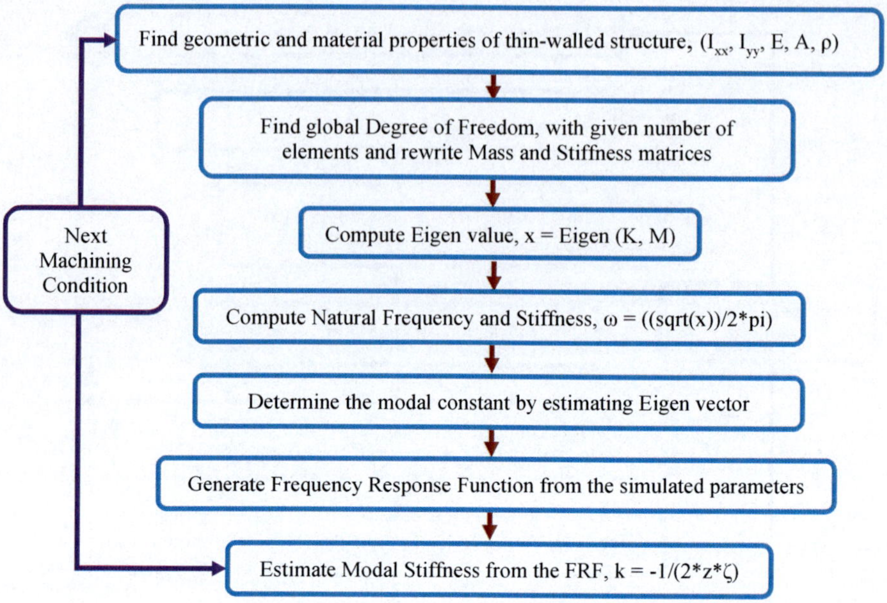

Fig. 7 Pseudocode for estimation of modal parameters from numerical simulation

is called as mass-normalized mode shape. The mode shape matrix for the system is $\{\Psi\}$ which is normalized by:

$$\{\Phi\}_r = \frac{1}{\sqrt{m_r}}\{\Psi\}_r, \quad r = 1, 2, \ldots.., n \tag{9}$$

where r is the total count of modes and m_r is the decoupled modal mass matrix, given by:

$$\{\Psi\}_r^T[M]\{\Psi\}_r = m_r \tag{10}$$

Here, in Eq. (6), $\{\Phi\}_r$ is the mass-normalized mode shape of the system for the rth mode. In matrix form for the rth mode, the orthogonality of the system can be expressed as:

$$\{\Phi\}^T[M]\{\Phi\} = [I] \tag{11}$$

$$\{\Phi\}^T[K]\{\Phi\} = \left[\omega_r^2\right] \tag{12}$$

This normalization is done to express FRFs of the system in terms of natural frequency and mode shapes. The matrices $[\omega_r{}^2]$ and $[\Phi]$ are derived using global matrices (mass matrix and stiffness matrix), and the response of the system contains the required FRFs.

Using the orthogonality equations [Eqs. (11) and (12)], the receptance FRF matrix given by Eq. (13) can be re-written as shown in Eq. (14):

$$[\alpha(\omega)] = \left([K] - \omega^2[M]\right)^{-1} \tag{13}$$

$$[\Phi]^T\left[[K] - \omega^2[M]\right][\Phi] = [\Phi]^T[\alpha(\omega)]^{-1}[\Phi] \tag{14}$$

$$[\alpha(\omega)] = [\Phi]\left[\omega_r^2 - \omega^2\right]^{-1}[\Phi]^T \tag{15}$$

Equation for a single receptance frequency response function $\alpha_{jk}(\omega)$ can be expressed by rewriting Eq. (12) as:

$$\alpha_{jk}(\omega) = \frac{\Phi_{j1}\Phi_{k1}}{\omega_1^2 - \omega^2} + \frac{\Phi_{j2}\Phi_{k2}}{\omega_2^2 - \omega^2} + \ldots + \frac{\Phi_{jn}\Phi_{kn}}{\omega_n^2 - \omega^2} \tag{16}$$

Alternatively,

$$\alpha_{jk}(\omega) = \left\{\dot{r}A_{jk}\right\}\left\{\frac{1}{\omega_r^2 - \omega^2}\right\} \tag{17}$$

Similarly, for damped system, Eq. (17) can be re-written as

$$\alpha_{jk}(\omega) = \frac{1}{\left(\omega_r - \omega^2\right) + 2 * \zeta * \omega_r * \omega} \tag{18}$$

where $\{{}_rA_{jk}\}$ is the product of jth and kth elements in the rth mode shape, and it is called modal constant. Every column of modal matrix will have its own modal constant. A constant damping of 0.3% had been used for all modes. The generated complex FRFs are splitted into its corresponding real and imaginary parts and are plotted against the frequency to get the modal frequency and modal stiffness of the thin-walled structure. The procedure is followed again for all possible machining conditions to predict modal parameters.

5.3 Experimental Modal Analysis for Dynamics of Thin-Walled Ti6Al4V

Dynamics of thin-walled Ti6Al4V has been estimated using modal analysis method. As the workpiece is stationary, impact and response have been collected at the same point of impact at different locations along the tip of thin-walled structure (Fig. 4b). Mean of the captured signal has been removed from both impact and response to

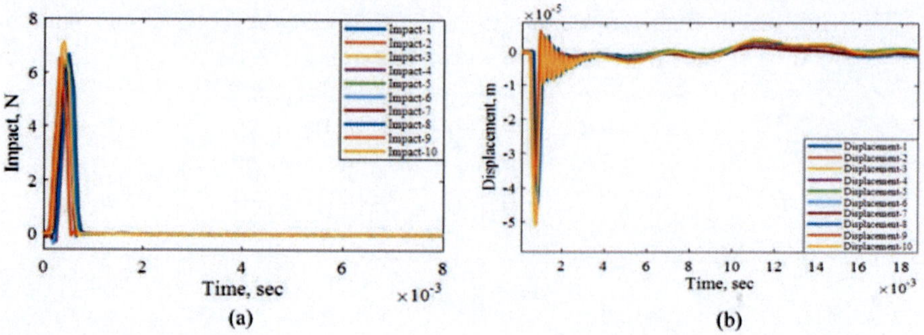

Fig. 8 Impact excitation and its response for thin-walled Ti6Al4V workpiece; **a** impact in N, **b** displacement in m

avoid the DC in the signal. The estimation of the modal parameters from FRFs of thin-walled Ti6Al4V workpiece has been carried out using peak-picking method. A thin-walled 'T'-shaped Ti6Al4V (Fig. 5) has been used in the current work. The excitation to the thin-walled workpiece is given by impact hammer (Dytran 5800SL) at three different locations along its length (Fig. 4b). The maximum frequency bandwidth of Dytran 5800SL impact hammer is 40 kHz. An average of ten measurements has been taken for estimation of FRF for static thin wall. The displacement of the excited thin wall has been measured using Laser Displacement Sensor (OptoNCDT, ILD2300 from Micro-Epsilon) in the same direction as that of impact with spot diameter of 23 μm. The displacement has been measured at three different locations to estimate the FRF at the tip of the thin-walled Ti6Al4V. The responses for all the impacts (Fig. 8a) are given in Fig. 8b. Sufficient time has been given after the impact, such that the displacement decays to zero completely before the next impact is given, this is evident in Fig. 8b.

The FRF ($\Phi(\omega)$) has been obtained by impacting the thin wall near the tip portion, and response is collected at the same point. The FRF ($\Phi(\omega)$) at the tool-tip is obtained by:

$$\Phi(\omega) = \frac{X(\omega)}{F(\omega)} \tag{19}$$

where $F(\omega)$ is the impact force at the point of impact (Fig. 4b) and $X(\omega)$ is the recorded response at the same point (Fig. 4b). To get the modal parameters from obtained FRFs, peak-picking method was used. Peak-picking method of extracting modal parameters from estimated tool-tip FRF is described in Fig. 9. The damping ratio is obtained by using the relation:

$$\zeta_i = \frac{\omega_b - \omega_a}{2 * \omega_{n,i}} \tag{20}$$

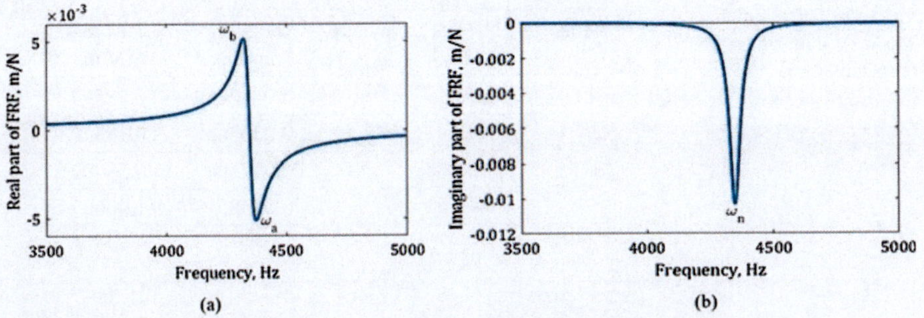

Fig. 9 Modal parameter extraction from FRFs by peak-picking method

where ω_b and ω_a are the frequencies identified in the real plot (Fig. 9a) for the corresponding $\omega_{n,i}$ in the imaginary plot (Fig. 9b). The dominant peak in the imaginary plot of FRF is the natural frequency, where phase changes from 0° to 180°. Once the damping ratio is obtained, the dynamic stiffness can be calculated by using the equation below:

$$k = \frac{-1}{2 * \zeta_i * Z} \tag{21}$$

where Z is the magnitude of $\omega_{n,i}$ as measured in the imaginary plot of FRF (Fig. 9b).

The estimated modal parameters (natural frequency, damping ratio and stiffness) using peak-picking method (Table 4) have been used to reconstruct the FRFs. The comparison of real and imaginary parts of reconstructed FRF with experimentally measured FRF is shown in Figs. 10a, b. The reconstructed FRF for machining condition-1 shows a good agreement with the experimentally measured FRFs except at mode-1 (Fig. 10a). The observed deviation of the measured FRF from the reconstructed FRF at mode-1 is because of the difference between simulation and experimentation. During simulation of the thin wall, to reduce the computational time and cost, the base part which was held inside a vise (see Fig. 11) rigidly during experimentation has not been modelled. This change in modelling method has reduced the number of degree of freedom to be solved for, but in experimental condition as this is held inside a vise firmly, it contributed to the changes in mode-1. This estimated mode-1 from experimental modal analysis represents the shank mode and not the thin-walled tip mode. Hence, the second mode in simulation is considered as first mode (tip mode) for the thin-walled workpiece.

Table 4 Thin-walled Ti6Al4V dynamics (experimental)

Mode	Natural frequency, ω, Hz	Damping constant, ζ	Stiffness, k, MN/m
1	4453.13	0.0482459	0.18394
2	6445.31	0.0363636	0.19379
3	7421.88	0.0078949	0.53002
4	7890.62	0.0148518	1.46609

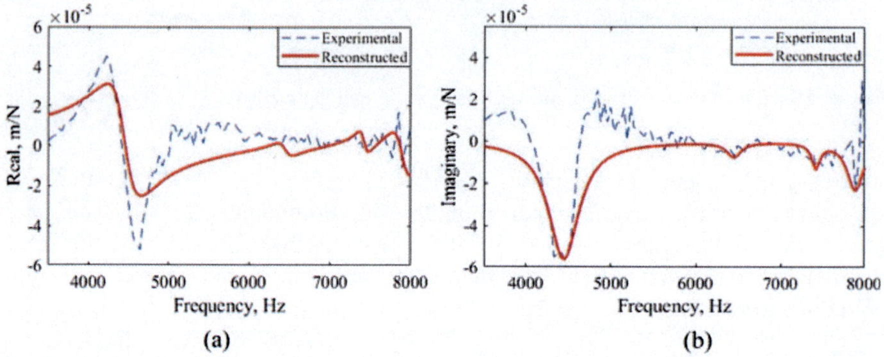

Fig. 10 Experimental and reconstructed real and imaginary parts of FRF for thin-walled Ti6Al4V; **a** real part, **b** imaginary part

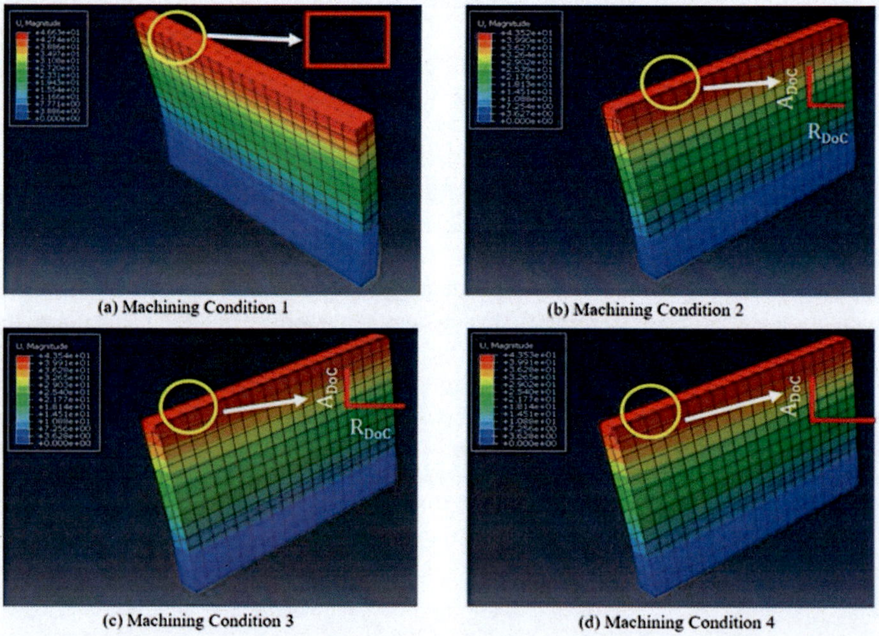

Fig. 11 Numerical simulation of flexible thin-walled Ti6Al4V at different machining conditions

6 Results and Discussion

6.1 Prediction of Modal Parameters for Different Machining Conditions

The simulation results indicate that the modal stiffness of the thin-walled structure decreases with increase in radial depth of cut (Table 5). The percentage reduction in stiffness from initial machining condition (condition-1) to final machining condition (condition-4) is 27% as shown in Fig. 13. From Eq. (22), it is observed that the stiffness of the beam is directly proportional to moment of inertia (MoI) of the beam for a given cross section. With increment in radial depth of cut, the moment of inertia of the thin-walled structure decreases (Table 5), thereby stiffness of the thin-walled structure reduces (Fig. 13).

$$k = \frac{3EI}{l^3} \tag{22}$$

where k is stiffness of the structure, E is Young's modulus, I is the MoI, and l is length of the thin-walled workpiece. The Mode-1 (bending mode) contour plots of frequency analysis performed in ABAQUS® for different machining condition are shown in Fig. 11. As material removal increases with the increase in radial depth of cut, the natural frequency decreases as the deflection of the thin-walled structure increases. The decreasing trend in natural frequency (Fig. 12) is observed except at the machining condition-1, because there is no engagement of cutting tool and work as radial depth of cut is zero. The frequency at machining condition-1 is the natural frequency of the thin-walled material without removal of any material. From machining condition-2 to machining condition-3, i.e. as radial depth of cut increases from 40 to 90 μm, the natural frequency drops by 5.08% and from machining condition-3 to machining condition-4; i.e. as radial depth of cut increases from 90 to 140 μm, the natural frequency drops by 5.26%. An overall decrement of

Table 5 Prediction of modal parameters through numerical simulation for different machining conditions

Parameters	Machining condition-1	Machining condition-2	Machining condition-3	Machining condition-4
Moment of inertia, $\times 10^{-12}$ m^4	5.625	5.1868	4.6720	4.1924
Cross-sectional area, $\times 10^{-6}$ m^2	30	20.998	20.99955	20.993
Modal stiffness, k, MN/m	0.17936	0.16539	0.14897	0.13368
Modal frequency, ω, Hz	6132.5	7038.8	6680.8	6329.0

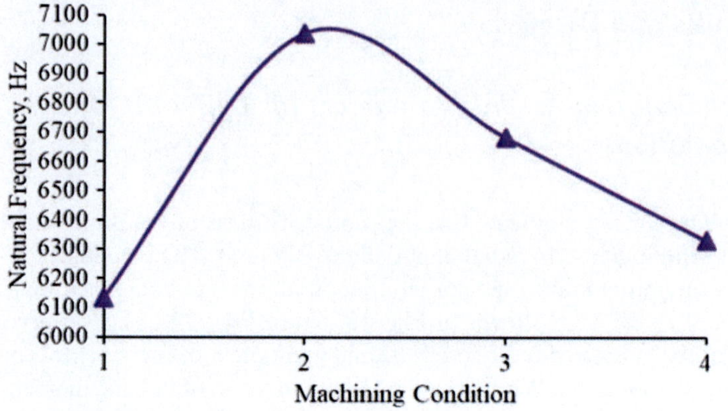

Fig. 12 Variation of natural frequency with different machining conditions

Table 6 Prediction of modal parameters through numerical simulation and validation by experimental modal analysis for 0 μm-radial depth of cut

Parameters	Simulation		Validation
	ABAQUS	MATLAB	Experimental modal analysis
Natural frequency, ω, Hz	6006	6132.5	6445.31
Stiffness, k, MN/m	0.676	0.17936	0.19379

10.08% in natural frequency is observed when radial depth of cut increases from 40 to 140 μm as seen in Fig. 12. Table 6 gives the values of modal parameters predicted by simulation and validated by experimental modal analysis (EMA). The predicted and simulated values of modal parameters are in good agreement with each other. The percentage error in prediction of natural frequency is 5.1%, while the percentage error in prediction of stiffness is 8.04 % for the first mode. Thus, the proposed numerical analysis for modal parameter prediction can be used as an alternate low-cost, time-saving technique for manufacturing thin-walled structures (Fig. 14).

6.2 Effect of Radial Depth of Cut on Stability

By solving the characteristic equation given in Eq. (3) for different spindle speeds, stable depth of cut is obtained as a function of spindle speed. The obtained stable depth of cut is plotted against spindle speed to obtain the stability lobe diagram for the given cutting conditions. Figure 15 shows the generated stability lobes for different machining conditions. The flexibility-dependent dynamics decreases the stable depth of cut owing to reduction in dynamic stiffness (Fig. 14). This reduction in stable depth of cut is due to the changing stiffness of the thin-walled workpiece with the increase in radial depth of cut. For machining condition-1, (Fig. 15a) at

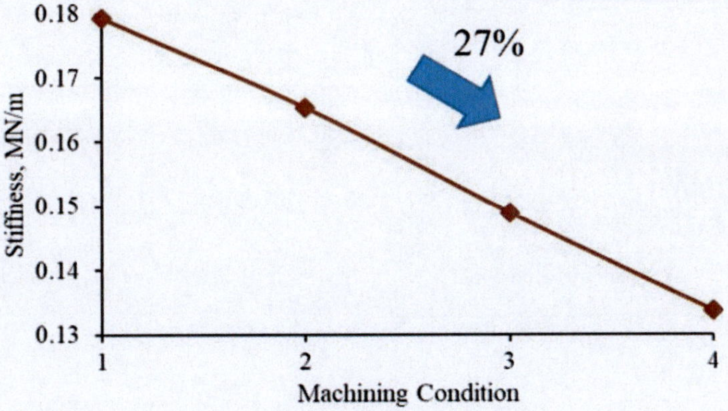

Fig. 13 Variation of stiffness with different machining conditions

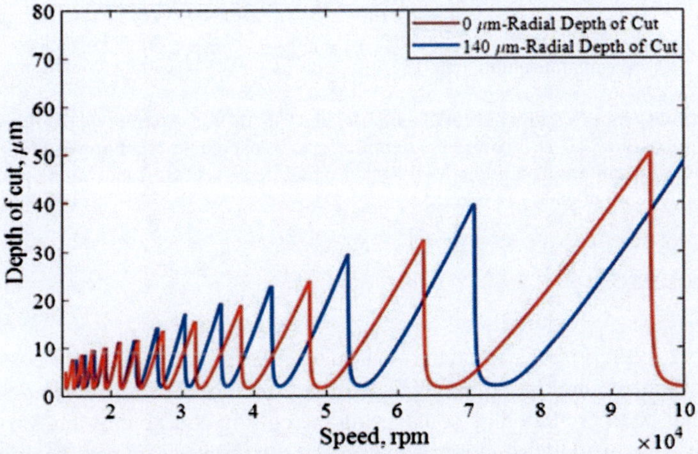

Fig. 14 Influence of radial depth of cut on stability for different machining conditions

60,000 rpm, the critical value of depth of cut is 40 μm, and for machining condition-4, (Fig. 15d) at 60,000 rpm, the critical value of depth of cut is 15 μm. By increasing radial depth of cut from 0 to 140 μm, the stable depth of cut reduces by 62.50% at 60,000 rpm. Similarly, at 80,000 rpm, the reduction in stable depth of cut is 66.67% from 24 to 8 μm at machining condition-1 (Fig. 15a) and machining condition-4 (Fig. 15d), respectively.

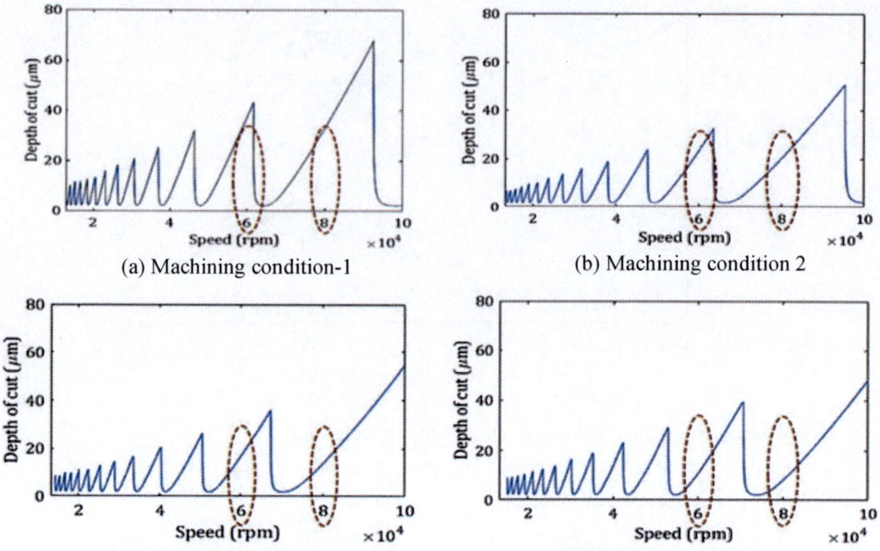

Fig. 15 Stability lobe diagram for flexible thin-walled Ti6Al4V at different machining conditions; **a** machining condition-1, at 0 μm radial depth of cut, **b** machining condition-2, at 40 μm radial depth of cut, **c** machining condition-3, at 90 μm radial depth of cut, **d** machining condition-4 at 140 μm radial depth of cut

7 Conclusion

The present work proposes numerical simulation as an alternative for cost and time intensive experimental modal analysis for manufacturing of thin-walled structures. Stability has been predicted by considering the cutting coefficients and varying stiffness of the thin-walled structure. The predicted modal parameters have been validated by performing experimental modal analysis on flexible thin-walled Ti6Al4V for static condition. The following are the conclusions from the present work:

- The natural frequency predicted from the simulated modal analysis of thin-walled Ti6Al4V workpiece has the deviation of 5.1% compared to natural frequency determined by experimental modal analysis for the first mode.
- The stiffness predicted from the simulated modal analysis of thin-walled Ti6Al4V workpiece has the deviation of 8.04% compared to stiffness determined by experimental modal analysis for the first mode.
- There in an increase in the natural frequency from no cut in thin-walled workpiece to just start of the material removal for radial depth of cut 40 μm. However, there is a continuous decrease of natural frequency with an increase in radial depth of cut.
- The natural frequency decreases by 11.2% by increasing the radial depth of cut from 40 to 140 μm.

- With increase in radial depth of cut from 0 to 140 μm, stiffness of the thin-walled workpiece has reduced by 27% compared to stiffness with zero removal of the material from the thin-walled workpiece.
- For 27% reduction in stiffness, the stable depth of cut drops by 62.50% at 60,000 rpm. This shows the necessity to incorporate the flexibility of the workpiece into the model while predicting the stable process parameters.
- Calculating modal parameters of thin wall machining at all machining conditions is impractical. Hence, finite element (FE) simulation proves to be a cost effective method.
- The stability lobe has also been developed analytically showing the effect of workpiece flexibility on stable depth of cut.
- There can be significant increase in the cutting tool life and the material removal rate for the chosen set of stable process parameter, as the developed lobe diagram incorporates the effect of varying stiffness.

References

1. Li Y, Cheng X, Zheng G, Yan J, Liu H, Li X (2022) Dynamic modeling and in-process parametric compensation for fabricating micro straight thin walls by micromilling. J Manuf Res Tech 18:2480–2493. ISSN: 2238-7854
2. Wang F, Cheng X, Zheng GM, Yang XH, Guo QJ, Sun QL (2019) Study of micromilling parameters and processes for thin wall fabrications. J Preci Engg 56:246–254. ISSN: 0141-6359
3. Ismail F, Ziaei R (2002) Chatter suppression in five-axis machining of flexible parts. Int J Mach Tool Manufac 42(1):115–122
4. Eksioglu C (2011) Mechanics and dynamics of thin-wall machining. MASc, Thesis. The University of British Columbia, Vancouver, BC, Canada
5. Fischer A, Eberhard P, Ambrósio J (2013) Parametric flexible multibody model for material removal during turning. J Comp Non-lin Dyna 9(1):011007
6. Gerasimenko A, Guskov M, Duchemin J, Lorong P, Gouskov A (2015) Variable compliance-related aspects of chatter in turning thin-walled tubular parts. Pro CIRP 31:58–63
7. Lorong P, Larue A, Perez Duarte A (2011) Dynamic study of thin wall part turning. In: 13th CIRP international conference on mod mechanical operational advanced materials research, Sintra, Portugal, pp 591–599
8. Mehdi K, Rigal JF, Play D (2002) Dynamic behavior of a thin-walled cylindrical workpiece during the turning process, part 1: cutting process simulation. J Manuf Sci Eng 124(3):562–568
9. Stepan G, Kiss AK, Ghalamchi B, Sopanen J, Bachrathy D (2017) Chatter avoidance in cutting highly flexible workpieces. CIRP Annals 66(1):377–380
10. Urbikain G, López de Lacalle LN, Campa FJ, Fernández A, Elías A (2012) Stability prediction in straight turning of a flexible workpiece by collocation method. Int J Mach Tools Manuf 54–55:73–81
11. Budak E, Altintas Y (1995) Modeling and avoidance of static form errors in peripheral milling of plates. Int J Mach Tools Manuf 35(3):459–476
12. Elbestawi MA, Sagherian R (1991) Dynamic modeling for the prediction of surface errors in the milling of thin-walled sections. J Mat Pro Tech 25(2):215–228
13. Kline WA, DeVor RE, Shareef IA (1982) The prediction of surface accuracy in end milling. J Eng Ind 104(3):272–278
14. Tsai JS, Liao CL (1999) Finite element modeling of static surface errors in the peripheral milling of thin-walled workpieces. J Mat Pro Tech 94(2):235–246

15. Tsai MP, Tsai NC, Yeh CW (2016) On milling of thin-wall conical and tubular workpieces. Mech Sys Sig Pro 72–73:395–408
16. Wan M, Zhang W, Qiu K, Gao T, Yang Y (2005) Numerical prediction of static form errors in peripheral milling of thin-walled workpieces with irregular meshes. J Manuf Sci Eng 127(1):13–22
17. Koike Y, Matsubara A, Yamaji I (2013) Design method of material removal process for minimizing workpiece displacement at cutting point. CIRP Ann Manuf Tech 62(1):419–422
18. Bravo U, Altuzarra O, López de Lacalle LN, Sánchez JA, Campa FJ (2005) Stability limits of milling considering the flexibility of the workpiece and the machine. Int J Mach Tools Manuf 45(15):1669–1680
19. Thevenot V, Arnaud L, Dessein G, Cazenave–Larroche G (2006) Influence of material removal on the dynamic behavior of thin-walled structures in peripheral milling. Mach Sci Tech 10(3):275–287
20. Mane I, Gagnol V, Bouzgarrou BC, Ray P (2008) Stability-based spindle speed control during flexible workpiece high-speed milling. Int J Mach Tools Manuf 48(2):184–194
21. Arnaud L, Gonzalo O, Seguy S, Jauregi H, Peigné G (2011) Simulation of low rigidity part machining applied to thin-walled structures. Int J Adv Manuf Tech 54(5):479–488
22. Song Q, Ai X, Tang W (2011) Prediction of simultaneous dynamic stability limit of time–variable parameters system in thin-walled workpiece high-speed milling processes. Int J Adv Manuf Tech 55(9):883–889
23. Liu Y, Wu B, Ma J, Zhang D (2016) Chatter identification of the milling process considering dynamics of the thin-walled workpiece. Int J Adv Manuf Tech 89(5–8):1765–1773
24. Seguy S, Dessein G, Arnaud L (2008) Surface roughness variation of thin wall milling, related to modal interactions. Int J Mach Tools Manuf 48(3–4):261–274
25. Adetoro OB, Sim WM, Wen PH (2010) An improved prediction of stability lobes using nonlinear thin wall dynamics. J Mat Pro Tech 210(6):969–979
26. Kersting P, Biermann D (2014) Modeling techniques for simulating workpiece deflections in NC milling. CIRP J Manuf Sci Tech 7(1):48–54
27. Altintas Y (2000) Manufacturing automation: principles of metal cutting and machine tool vibrations. Cambridge University Press, New York

Corrosion Behavior and Its Effect on Mechanical Properties of ER70S-6 Cladding on AA 6061-T6 Alloy Using a Cold Metal Transfer Process

Bappa Das, Biranchi Narayan Panda, and Uday Shanker Dixit

1 Introduction

Aluminum (Al) and its alloys have low density and good ductility, making it one of the widely used materials in the aircraft and automobiles industries. However, due to their low hardness and wear resistance, it is not suitable for certain sensitive applications. Many scientists are trying to improve the surface properties of Al by cladding/coating and surface alloying with different metallic and ceramic powders. Generally, Al has good corrosion resistance; however, in the acidic medium, the protective layer breaks and corrosion starts. Corrosion has a profound effect on the strength and stability of structures and has received much attention in engineering. Corrosion is a serious problem in components of gas and diesel engines, aircraft components and heat exchangers. The salt and acid solutions are harsh working environment that corrode metals. Chemicals like H_2SO_4 attack metals like copper (Cu), stainless steel and their alloys. Fouad et al. [1] reported the corrosion behavior of Cu and stainless steel in an acidic medium of H_2SO_4. The rate of corrosion was high for copper as compared to stainless steel. Moorthy et al. [2] reported corrosion of mild steel, aluminum, copper, and other metals because of base coolants used in automobiles during a 14-day corrosion test. One of the easiest and most direct ways to assess corrosion of any kind is weight loss measurement in a corrosion immersion test.

B. Das (✉) · B. N. Panda · U. S. Dixit
Indian Institute of Technology Guwahati, Guwahati, Assam, India
e-mail: bappadas@iitg.ac.in

B. N. Panda
e-mail: pandabiranchi@iitg.ac.in

U. S. Dixit
e-mail: uday@iitg.ac.in

S. N. Joshi et al. (eds.), *Low Cost Manufacturing Technologies*,
https://doi.org/10.1007/978-981-19-8452-5_2

This study investigated corrosion behavior of ER70S-6 cladding on Al base plate using cold metal transfer (CMT) technology. CMT is a relatively new and cost-effective welding and cladding method with limited heat input and is suitable for a variety of materials. Developed by Fronius of Austria in 2004, CMT is an improved gas metal arc welding (GMAW) technique based on a short circuit transfer mechanism that deposits materials with enhanced weld bead aesthetic and low thermal input [3]. The controlled feeding of the electrode in CMT technique reduces the heat used for weld beads, resulting in less distortion and increased productivity [4]. In a CMT process, wire is used as a feedstock material, and robotic arm is used for depositing the metal, which can move on a complex route [5]. Cladding by the CMT process consists of depositing material on the substrate to enhance the targeted microstructural and mechanical properties. Heat input during deposition as well as heat transfer after deposition are the main factor influencing the properties of the cladded or coated material [6].

In the present study, an Fe-based electrode was cladded on Al substrate using CMT process. Filler rod ER70S-6 was chosen due to its low-cost and high-deposition accuracy. The primary objective of the present study is to analyze the corrosion behavior of the Fe-based cladding on the Al substrate. It is an extension of the previous work of the authors [7], focusing on the surface roughness, microhardness, and wear before and after corrosion.

2 Experimental Procedure

The experiments were conducted on Al plate with dimension of 180 mm × 100 mm × 6 mm. On AA 6061-T6 alloy substrate, cladding of ER70S-6 was accomplished. The weld metal composition is provided in [7] and ESAB, India [8]. The cladded layer was deposited with a Fronius CMT TPS400i machine. The shielding gas used was argon at 20 l/min constant flow rate. The parameters for CMT process were 100 A current, 400 mm/min welding speed, 10 m/mm wire feed speed, 300 s interlayer dwell time, 10 V voltage, 8 mm/s welding torch speed, and 5 mm standoff distance. The final height of the cladding achieved was 3 mm in single pass. Figure 1 shows the CMT setup and extracted characterization sample after milling. The cladded layer was milled to a depth of approximately 1 mm to remove the uneven surface prior to sample extraction for characterization.

For microstructural, corrosion, and mechanical characterization, the sample was sectioned through electric discharge machining after milling. The sample was polished on various grits (sizes: 80, 400, 800, 1000, 1500, and 2000) of silicon carbide emery papers followed by cloth and diamond polishing. For corrosion test, a sample of CMT cladded alloy was tested in a solution of 2.0% H_2SO_4 for 200 h according to ASTM standard: G31. The CMT cladded sample was mounted in a corrosion-resistant cold mount with top surface area of 100 mm^2 exposed to acidic solution at room temperature. To calculate the weight loss, the corroded sample was weighed with an accuracy of 0.1 mg. A non-contact optical surface profilometer was

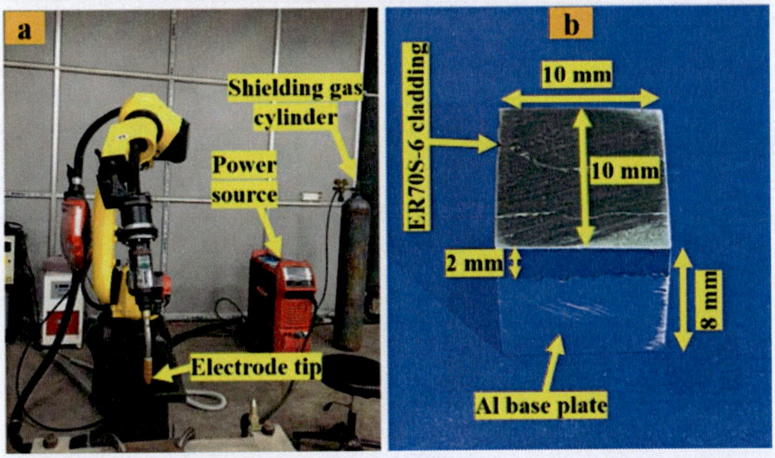

Fig. 1 Cladded layer developed using CMT technique: **a** setup and **b** extracted characterization sample after milling

used to measure the corrosion pits height and surface roughness. An upright optical microscope and a field emission scanning electron microscope (FESEM) were used to investigate the sample's morphology. A microhardness tester was used to test the Vickers microhardness with 20 s as dwelling period while applying a 200 gf load. The pin-on-disk wear test was conducted on a tribometer for 30 min with 20 N load at 350 rpm. Cladded material in the form of 10 mm diameter pin traversed on 70 mm track diameter on hardened steel disk according to ASTM standard: G99. The detailed specifications of machines are provided in [7].

3 Results and Discussion

This section discusses a few findings. First, the microstructure is described. Next, the surface roughness due to corrosion is discussed. Finally, the microhardness and wear results before and after the corrosion test are described.

3.1 Corrosion Mechanism and Microstructural Study

Understanding the mechanism of corrosion is essential to comprehending the physics of corrosion. The mechanisms of steel corrosion in aqueous solution is reported by a few authors [9–11]. The corrosion mechanisms of steel in presence of oxygen are as follows:

Anodic reaction:

$$Fe \rightarrow Fe^{2+} + 2e^- \tag{1}$$

Cathodic reaction:

$$O_2 + 4e^- + 2H_2O \rightarrow 4OH^- \tag{2}$$

$$Fe^{2+} + 4OH^- \rightarrow Fe(OH)_2 \tag{3}$$

$$4Fe(OH)_2 + O_2 \rightarrow 2Fe_2O_3 + 4H_2O \tag{4}$$

Panossian et al. [12] reported the corrosion of steel in H_2SO_4 medium in absence of oxygen. Steel is immediately attacked by the creation of hydrogen gas and ferrous ions when it comes into contact with a dilute H_2SO_4 medium, as follows:

Anodic reaction: As per Eq. (1)

Cathodic reaction:

$$2H^+ + 2e^- \rightarrow H_2 \tag{5}$$

Therefore, the global corrosion reaction can be written as follows:

$$Fe + 2H^+ \rightarrow Fe^{2+} + H_2 \tag{6}$$

The composition of the steel, particularly the amount of carbon, has a significant impact on the rate of corrosion of carbon steels in diluted acids. As a result, the oxidation reaction ($Fe \rightarrow Fe^{2+} + 2e^-$) occurs in the ferrite phase, whereas the cathodic reduction reaction producing H_2 occurs at cementite phase. When the steel comes in contact with the H_2SO_4 medium, the latter reduced to form H_2 and Fe oxidized to form ferrous sulfate ($FeSO_4$), as follows:

$$Fe + H_2SO_4 \rightarrow 2FeSO_4 + H_2 \tag{7}$$

The $FeSO_4$ adheres to the steel surface and acts as a barrier. This $FeSO_4$ layer protects the steel surface from further corrosion attacks in the acid medium during a long-term process. As a result, the maintenance of the $FeSO_4$ layer determines the durability of steel products.

The corrosion immersion test should run for a longer duration of time. As per ASTM standard G31, test durations typically range from 48 to 168 h. Where the corrosion rate is less than 0.5 mm/year, the test should be conducted for at least 200 h. On alloys that generate passive films, short-time testing can also produce false findings. To allow for the breakdown of the passive film and a subsequent more rapid attack in borderline conditions, a long duration test may be necessary. As a result, tests that are done for extended periods of time are much more realistic than tests that are run for short duration. However, the corrosion should not advance to the point where the original specimen is perforated. Considering these factors, in the present work, the test was conducted for 200 h.

Figure 2 shows cross-sectional micrographs and EDS analyses of the cladded layer after the CMT process. The morphology depicts a well-defined and crack-free cladded layer deposited on Al base plate. The presence of different elements in the cladded layer and substrate region was confirmed by EDS analysis (Fig. 2c, d). The optical and FESEM images are shown in Fig. 3 to establish the interaction of H_2SO_4 with the metal surface in the cladding layer. The micrograph morphology shows the features of cladded surface before and after 200 h of corrosion in 2.0% H_2SO_4. The FESEM images reveal that the cladded surface after corrosion is covered with corrosion products and appears to be full of pits and cavities. The corrosion cracks are also visible in a few places in the corroded surface. The weight loss after the corrosion test was 0.089 gm. As the exposed area was 100 mm^2, this amounts to a weight loss of 8.9×10^{-4} g/mm^2.

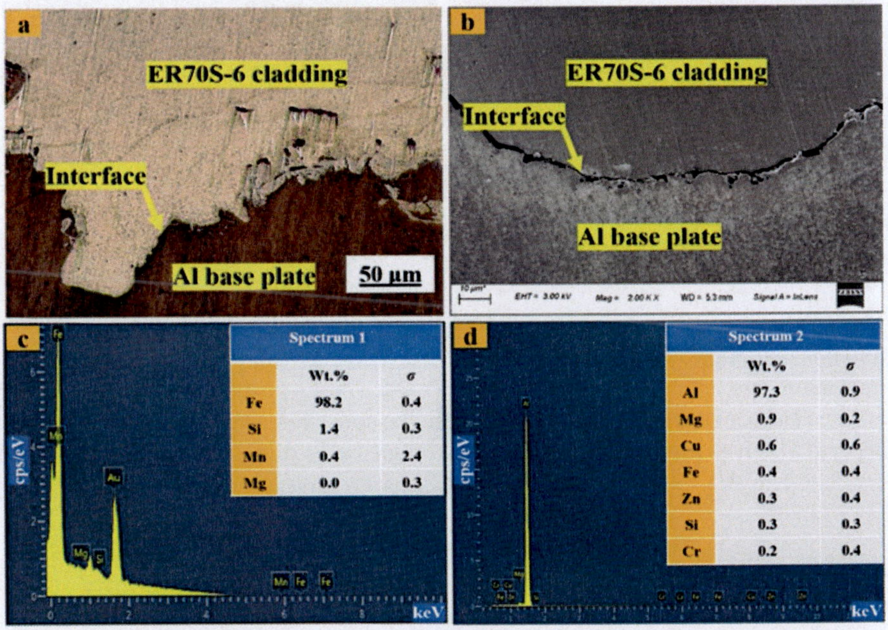

Fig. 2 Micrograph after cladding: **a** optical (X20), **b** FESEM (X2000); EDS elemental distribution of **c** cladded and **d** base plate region

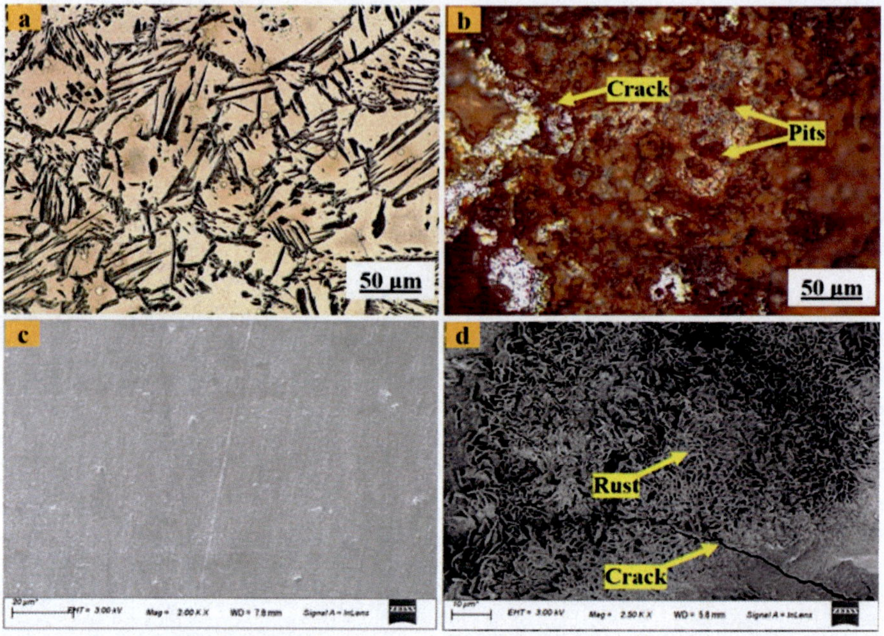

Fig. 3 Morphology of cladded layer: (1) optical micrograph (X20): **a** before corrosion and **b** after corrosion, (2) FESEM micrograph: **c** before corrosion (X2000) and **d** after corrosion (X2500)

3.2 Surface Roughness

The non-contact optical surface profilometer was used for measuring the surface roughness after the corrosion test. Figure 4 shows the 3D morphology after corrosion immersion test. The corrosion product iron oxide (Fe_2O_3) is observed in Fig. 4b. The average surface roughness value (R_a) before the corrosion test was 0.429 μm, which became 3.10 μm after corrosion test. Mean height before the corrosion was 0.391 μm and 5.46 μm after 200 h corrosion test.

3.3 Hardness Study

Figure 5 shows the average microhardness value of cladded layer before and after the corrosion test. Corrosion removes the hard layer of metal and reduces the hardness. The average microhardness value before corrosion was 237 HV, but after the corrosion test, the hardness reduced to 213 HV. The reduction in the microhardness value indicates that some depth of the corrosion-resistant alloy zone has been removed. This reduction attributes to the dissolution of metal ions and hard iron oxide layer

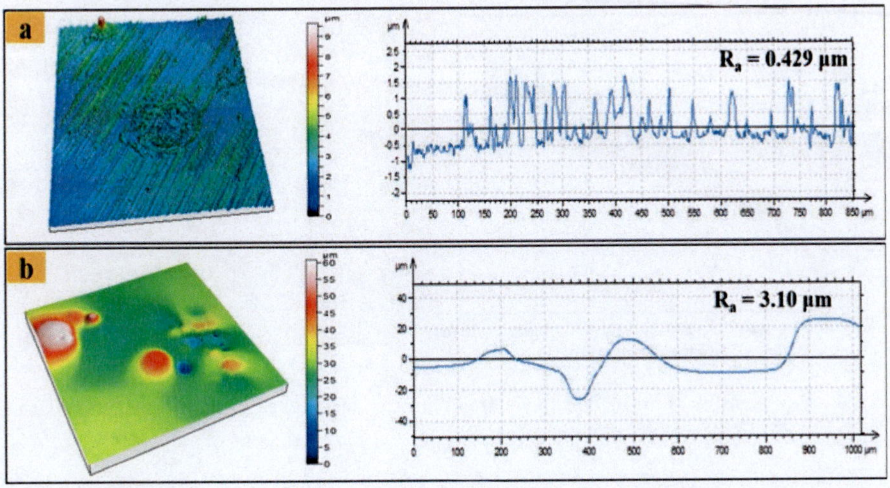

Fig. 4 3D images and corresponding roughness plot of cladded layer: **a** before corrosion test and **b** after corrosion test

in the corrosive medium [13]. The microhardness also decreased due to increased pitting corrosion, thus reducing the load-bearing area at certain places.

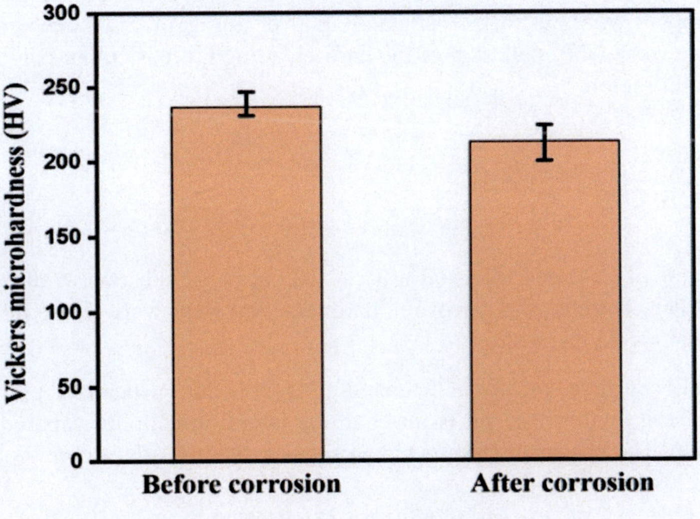

Fig. 5 Effect of corrosion on microhardness of the cladded layer

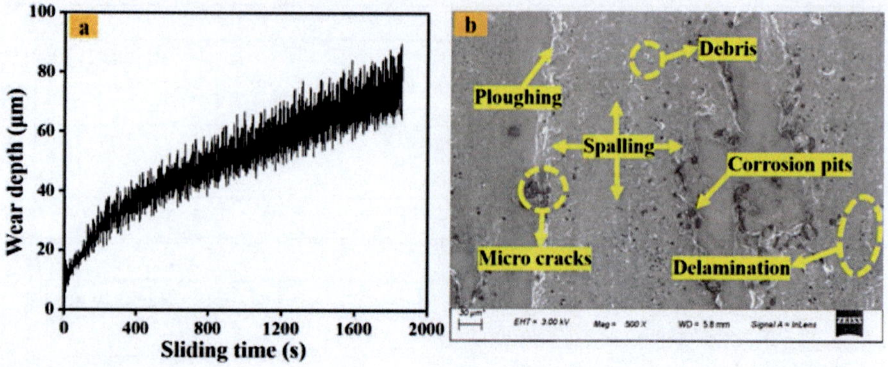

Fig. 6 Wear analysis of the cladded layer after corrosion test: **a** wear depth with sliding distance and **b** FESEM (X500) morphology

3.4 Wear Study

Figure 6 shows the wear data after the corrosion test. The average wear depth was 51.61 μm, which was 37 μm before corrosion test for the same test condition [7]. This 39.5% increase in wear depth after corrosion was due to the removal of hard particles from the upper most section of the cladding layer. The fluctuation of the wear depth data also increased due to the presence of corrosion pits in the cladding layer. The morphology of worn out surface of cladding layer in Fig. 6b also shows indications of abrasive wear. The spallation of the hard-cladding layer is comparatively spread over a larger region.

4 Conclusion

The cladding of Fe-based filler rod was carried out on Al substrate with CMT technology. After cladding, the corrosion, hardness, and wear were studied. Following are the salient conclusions:

- The major surface damage in the cladding layer is due to localized pitting corrosion. The growth of the pit is faster in the lateral direction compared to depth direction. The shapes of the pits were irregular. After 200 h of corrosion test, the weight loss was 8.9×10^{-4} g/mm^2.
- The average microhardness after the corrosion test reduced by 10% due to removal of hard materials and introduction of porosity, thus reducing the load-bearing area at certain places.
- The wear depth increased by 39.5% in the cladded region after 200 h corrosion test due to the presence of corrosion pits and removal of hard layer of metal.

Acknowledgements The Central Instrumentation Facility of the Indian Institute of Guwahati (IITG) provided the testing facility. Cold Metal Transfer Welding Machine used in this work was procured by IITG Technology Innovation and Development Foundation (IITGTI&DF), which has been set up at IIT Guwahati as a part of the National Mission on Interdisciplinary Cyber Physical Systems (NMICPS). IITGTI&DF is undertaking research, development, and training activities on Technologies for Under Water Exploration with the financial assistance from Department of Science and Technology, India through grant number DST/NMICPS/TIH12/IITG/2020. Authors gratefully acknowledge all the support provided for conducting the present research.

References

1. Fouad MA, Zewail TM, Amine NA, El-Tawail YA (2017) Comparison between corrosion behavior of copper and stainless steel 90° elbow and failure investigation of 90° copper elbow. J Inst Eng (India) Series C 98(2):141–145
2. Moorthy CV, Srinivas V (2016) Corrosion and heat transfer characteristics of water dispersed with carboxylate additives and multi-walled carbon nano tubes. J Inst Eng (India) Series C 97(4):569–577
3. Varghese P, Vetrivendan E, Dash MK, Ningshen S, Kamaraj M, Mudali UK (2019) Weld overlay coating of Inconel 617M on type 316 L stainless steel by cold metal transfer process. Surf Coat Technol 357:1004–1013
4. Tang X, Zhang S, Cui X, Zhang C, Liu Y, Zhang J (2020) Tribological and cavitation erosion behaviors of nickel-based and iron-based coatings deposited on AISI 304 stainless steel by cold metal transfer. J Mater Sci Technol 9:6665–6681
5. Selvi S, Vishvaksenan A, Rajasekar E (2018) Cold Metal Transfer (CMT) technology—an overview. Defence Technol 14(1):28–44
6. Rajeev GP, Kamaraj M, Bakshi RS (2017) Hardfacing of AISI H13 tool steel with stellite 21 alloy using cold metal transfer welding process. Surf Coat Technol 326:63–71
7. Das B, Panda BN, Dixit US (2022) Microstructure and mechanical properties of ER70S-6 alloy cladding on aluminum using a cold metal transfer process. J Mater Eng Perform. In press. https://doi.org/10.1007/s11665-022-06937-8
8. https://www.esabindia.com/in/en/products/filler-metals/mig-mag-wires-gmaw/mild-steel-wires/esab-mw1.cfm. Accessed on 5 May 2022
9. Kahyarian A, Schumaker A, Brown B, Nesic S (2017) Acidic corrosion of mild steel in the presence of acetic acid: mechanism and prediction. Electrochim Acta 258:639–652
10. https://www.suezwaterhandbook.com/water-and-generalities/corrosion-in-metal-and-concrete/protection-against-corrosion/cathodic-protection. Accessed on 2 May 2022
11. Hou X, Gao L, Cui Z, Yin J (2018) Corrosion and protection of metal in the seawater desalination. IOP Conf Series: Earth Environ Sci 108:022037
12. Panossian Z, Almeida NLD, Sousa RMFD, Pimenta GDS, Marques LBS (2012) Corrosion of carbon steel pipes and tanks by concentrated sulfuric acid: a review. Corros Sci 58:1–11
13. Jiru WG, Sankar MR, Dixit US (2019) Laser surface alloying of aluminum for improving acid corrosion resistance. J Inst Eng (India): Series C 100:481–492

An Experimental Investigation into CO_2 Laser-Based Processing of Boulder and Marble

Antash Kishore Sinha and Shrikrishna Nandkishor Joshi

1 Introduction

Laser-based rock drilling has emerged as an effective potential alternative to conventional drilling methods in the past two decades. It is a process of hole creation and enlargement in a material through light amplification by stimulated emission of radiation, abbreviated as LASER. It holds many advantages, such as green energy transmission, carbon emission reduction, non-contact material processing, and financial affordability over conventional methods [1, 2]. Laser rock drilling is a non-contact material interaction process. Hence, there is no requirement for any filler or machining tool material. Laser heat and power generation are exceptionally high compared to traditional drilling methods. This characteristic is helpful in the easier processing of materials such as rocks, composites, and ceramics, which are otherwise hard to process. The non-contact feature eliminates the requirement of rigs, drill bits, and casing in rock excavations. This can greatly aid in downsizing the material wastage and optimizing the process economy. However, due to certain challenges such as wellbore instability, nonlinear behavior of rocks, complex purging process, wellbore cleaning requirements, and difficulty in controlling downhole pressure and temperature, the field scale implementation of lasers in rock drilling is under rigorous research currently [2, 3]. When the rocks interact with lasers, they respond very volatilely. Rocks are exploited mainly for their mineral enrichments to meet global energy demands. Different rock formations are rich in different types of vital minerals depending upon their geographical and topological conditions. These minerals generally have lower melting and evaporation points than host rock material. Due to high power and heat generation in case of laser interaction, the minerals may evaporate before the beginning of rock material removal. This is challenging in terms

A. K. Sinha · S. N. Joshi (✉)
Department of Mechanical Engineering, Indian Institute of Technology Guwahati, Guwahati, Assam 781039, India
e-mail: snj@iitg.ac.in

S. N. Joshi et al. (eds.), *Low Cost Manufacturing Technologies*,
https://doi.org/10.1007/978-981-19-8452-5_3

of economy and process efficiency. In addition to this, with increasing depth, the downhole conditions start becoming adverse. Sustaining a good drilling rig structure is a prevailing challenge with both the traditional as well as the non-conventional potential drilling technologies.

The laser rock interaction process is influenced by many material properties and process parameters such as specific energy, rate of penetration, depth of penetration, laser power, scanning speed, scan shape, porosity, permeability, melting point, and so on [2, 4]. Several experimental works have been published to understand the response behavior of rocks to laser interaction. The central focus has been mainly on optimizing processing parameters, developing an effective numerical model, nonlinear finite element analysis, thermal zones analysis, and fracture mechanisms. The principal objective of most of these works has been to minimize specific energy consumption and maximize the rate and depth of penetration [5–7].

Rocks respond to lasers in a very unstable manner, accustomed to their inhomogeneity. Under the downhole condition, this inhomogeneity becomes vital, considering microcracks development aids in easy rock breaking. Li et al. [8] studied the influence of laser irradiation time and temperature gradient on granite's crack formation and failure mechanism. Based on the studies, a rock-breaking simulation model was developed employing discrete element method. Rock breaking was found to be a dynamic failure process and follows a crack-cracking mechanism of breakage. Deng et al. [9] pointed out the influence of different laser scan shapes on the strength and durability of granite. The rectangular scan shape was more capable of reducing granite's tensile and uniaxial compressive strength than the circular scan shape. Yan et al. [10] described the laser power and irradiation time as the most important variables influencing the laser-matter interaction. The depth of laser perforation could be increased by increasing the number of passes through the side blow or continuous zooming process.

Laser-based material interactions are challenging to control due to the different characteristics of lasers, such as high power and high coherency. Lasers have high intensity, due to which they raise the material temperature beyond vaporization when they interact with matter. Beyond melting and vaporization temperature, the material either melts or evaporates. In the melted or evaporated phase, the hydrodynamic stresses become a vital point of study to determine the material response. The hydrodynamic stresses alleviate laser power transmission and absorption by the workpiece surface. This explains the uncontrolled and nonlinear laser-matter interactions. Optical characteristic coefficients of irradiated matter, such as absorptivity, reflectivity, and scattering, play an essential role in governing the interaction process. Absorption plays a significant role in trapping the laser beam inside the pores of the material and increasing the heating effects, further leading to melting or vaporization [11, 12]. Schneider et al. [11] studied the influence of laser intensity variation on the absorptivity of aluminum, nickel, and steel. Irrespective of the nature of the material, the absorption was found to be as high as 80%. Lin et al. [12] emphasized the importance of micro-holes for absorption during laser drilling and developed an optical system for real-time precision measurement of their depths.

Achieving a high penetration depth in non-metallic hard rock structures is challenging because of many limitations. Graves et al. [4] enlisted one of the major challenges in having a higher rock penetration on laser interaction: finding an efficient high-power laser type. To date, the investigations have failed to establish an efficient laser with optimal laser power, reliability, durability, economic viability, and is environment-friendly. There is a requirement for enhancing well control in underbalanced and precarious downhole conditions. In the case of lasers, the power transmission is difficult over deeper depths, especially with the non-contact processing feature. Non-contact laser material processing becomes irrelevant at deeper depths due to unwanted material purges. O' Brien et al. [13] suggested that with the aid of proper casing, rigs, and traditional bit, if researchers can find a methodology to generate laser power in the downhole conditions at extreme pressure and temperature, then significant breakthroughs can be possibly achieved. Graves et al. [14] highlighted the role of mineral content, saturation medium, and material properties such as porosity for rocks in hindering the influence of laser from breaking the material. The dry and saturated conditions are vital in determining laser power absorption. Further, abundant mineral content presence is beneficial from the excavation aspect. However, it may sometimes lead to a complex purging phenomenon if it reduces the rock's porosity and hinders laser power absorption. Gahan et al. [15] found that vital zones of melting and spallation are formed during laser drilling of rocks. Specific energy is an important factor in determining the feasibility of the rock-drilling process. With the laser, the least specific energy was found prior to the melting point. However, it is not possible currently to heat the rock samples in a controlled manner till the desired critical temperature point before melting and stopping further laser power transmission and absorptions. Along with these factors, other hindrances from material and optical properties [16] also come into the picture. Hence, these challenges require significant attention from the scientific community to achieve a breakthrough for achieving a highly productive, economic, and environment-friendly drilling technology for a successful sustainable energy transition.

In this work, the influence of laser application on the boulder and marble material has been analyzed in terms of spallation, power usage, laser scan shape, and laser-processed zone shape. The possibility of finding penetration depth on laser-processed samples using an optical microscope has also been explored. The commonly available rock materials such as boulder material, construction purpose marble, and decoration purpose marble were selected for laser interaction. The irregularly shaped samples interacted with laser with the primary objective of developing a preliminary understanding of laser interaction with brittle rock materials. Parameters such as scanning speed and exposure time were varied in simultaneous experiments to understand the variation in material response. Further, the decoration marble has been used to analyze the potential use of an optical microscope for measuring the penetration depth in laboratory-scale samples. For a sample of decoration purpose marble, three different laser scan shapes were developed to analyze the influence on the volume and rate of material removal and depth of penetration.

2 Materials and Methods

Samples of irregularly shaped boulder material, construction marble, and decoration marble were selected for laser interaction. The dimensions of the samples were measured using Vernier Caliper and a 300-mm steel scale. Boulder material (Fig. 2) had a length of 88.93 mm, a width of 78.93 mm, and a thickness of 9 mm. Construction marble material (Fig. 3) was 37.39 mm in length, 21.31 mm in width, and 13.39 mm in thickness. The samples interacted with laser beams using a 2.5 kW CO_2 laser machine (Orion 3015 LVD) for different process parameter conditions in Tables 1 and 2. Primary interaction involved free alignment where a thorough penetration was attempted, as shown in Figs. 2 and 3. Secondary interactions were aimed at the development of a borehole of 5 mm diameter and depth at least at par with the thickness of the samples. For borehole creation, multiple passes of the laser beam were made on the boulder and marble sample to determine the number of passes needed for a better material removal. The experimental setup schematic is depicted in Fig. 1. The designs for these operations were made using CNC controls-CADMAN-L 3D software package. For analyzing the possibility of measuring the penetration depth using an optical microscope, a sample of marble used for decoration purposes was selected, having irregular dimensions with a length of 69 mm, a width of 64 mm, and a thickness of 18.5 mm. The decoration purpose marble interacted with a laser with three different scan shapes, circle, square, and rectangle, as depicted in Fig. 4. The dimensions of the scan shapes were a circle of 5 mm diameter, a square of 5 mm edges, a rectangle of 2.5 mm width, and a 5 mm height. Further, NIKON SMZ25 optical microscope was used to analyze the dimensions of processed zones and depth of penetration (Fig. 4 and Table 3).

Table 1 Process parameters for laser-boulder material interaction

Operation attempts	Power (in watt)	Scanning speed (in mm/min)	Standoff distance (in mm)	Spot diameter (in mm)	Pressure (in bar)	Number of passes of laser beam
Free alignment	1000	1200	1	0.5	20	3
Borehole cut 1	1000	1200	1	0.5	20	1
Borehole cut 2	1000	1200	1	0.5	20	2
Borehole cut 2	1500	1200	1	0.5	20	2

Table 2 Process parameters for laser-construction marble material interaction

Operation attempts	Power (in watt)	Scanning speed (in mm/min)	Standoff distance (in mm)	Spot diameter (in mm)	Pressure (in bar)	Exposure time (in sec)	Number of passes of laser beam
Free alignment cut 1	1500	600	1	0.5	20	1	1
Free alignment cut 2	1500	1200	1	0.5	20	1	1
Borehole operation 1	1500	1200	1	0.5	20	3	2
Borehole operation 2	1500	600	1	0.5	20	3	2

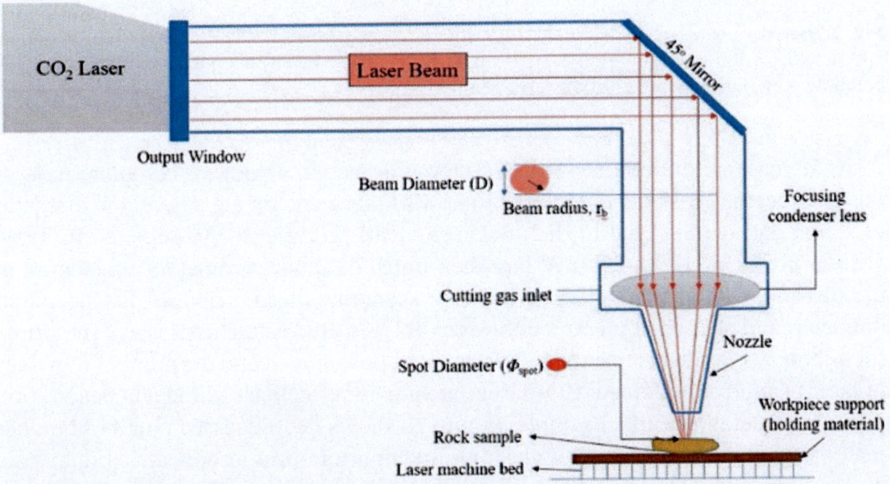

Fig. 1 Schematic of experimental setup [17]

3 Results and Discussion

The investigations presented in this work are preliminary with a hit and trial approach to analyze LASER's interaction with a non-metallic material like boulder and marble. The prime objective is to determine the feasibility of employing LASER in further investigations of potential rock fossil fields such as limestone, sandstone, and granite, spreading over the North-Eastern Indian territory.

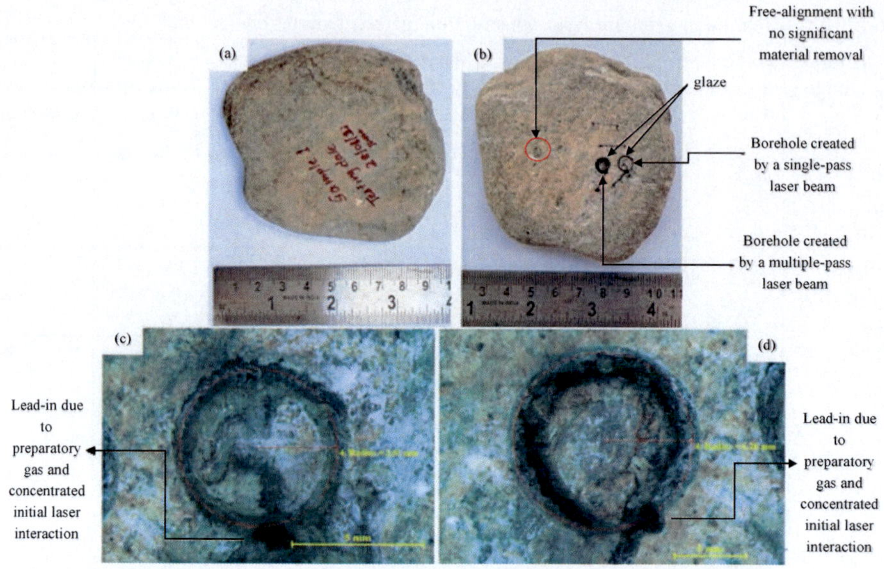

Fig. 2 Boulder material; **a** before LASER interactions, **b** after LASER interactions, **c** optical micrograph of the borehole with a single pass of the laser beam, **d** optical micrograph of the borehole with multiple passes of the laser beam

In the case of the boulder material, the aim was to achieve a complete pass of the laser beam across the thickness of the workpiece and create a borehole diameter of 5 mm and depth equal to the thickness of the workpiece. As depicted in Table 1, laser power of 1000–1500 W has been utilized. Other parameters considered in the investigation of boulder material were scanning speed, standoff distance, spot diameter, and assist gas pressure. Nitrogen (N_2) is used as the assist gas. The prime focus here was to determine the requirement of power level and the number of passes of laser beam from a 2.5 kW CO_2 laser machine for effective drilling and penetration into a non-metallic boulder sample. Figure 2b shows the processed zones of boulder material after laser interaction with single and multiple-pass laser beams. Figures 2c, d show the optical micrographs of boreholes with a single-pass laser beam and multi-pass laser beam interaction, respectively.

The first operation on boulder material was a free alignment operation with laser power of 1000 W at the location shown in Fig. 2b. The laser beam was passed thrice at this location. However, there was no success in having a suitable ablation even with multiple laser beam passes. Further, the creation of a borehole was attempted, as shown in Fig. 2b, for single as well as multiple passes of the laser beam at the depicted locations. For a single pass, 1000 W of laser power was used. The material response was unsatisfactory, as evident from the optical micrograph shown in Fig. 2c. The target was to achieve a borehole diameter of 5 mm, i.e., radius of 2.5 mm and an effective penetration depth equivalent to the thickness of the sample, i.e., 9 mm. With single pass, a borehole of a radius of 3.01 mm was created. The depth of penetration

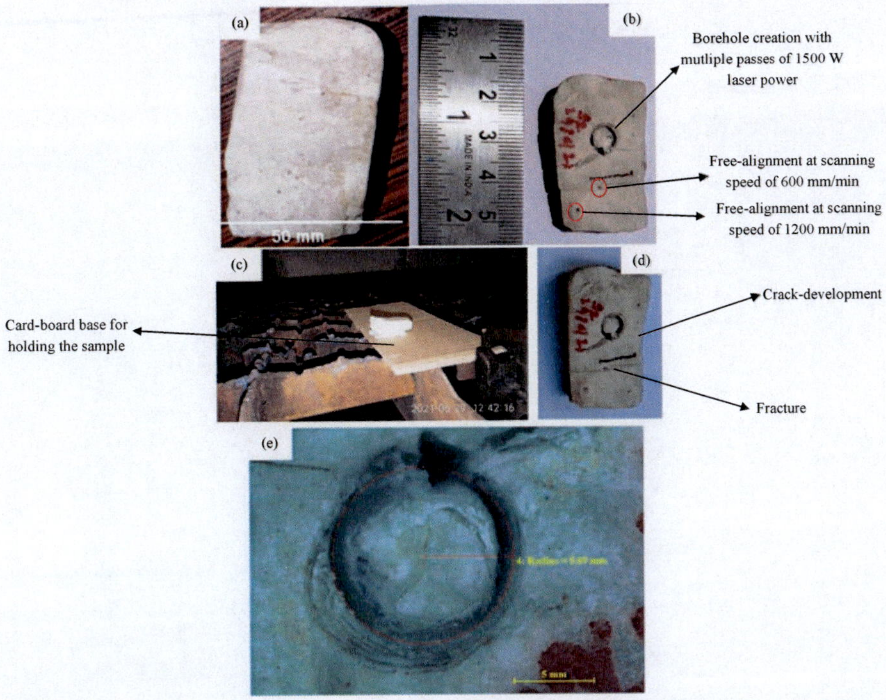

Fig. 3 Construction marble material; **a** before laser interaction (scaling done using ImageJ software), **b** after lasing action with the laser-processed zones of free alignments and borehole, **c** sample kept on laser machine bed, with a cardboard material used as workpiece holding base along with clamps, **d** cracks and fracture zones, **e** optical micrograph depicting borehole of 11.38 mm diameter

could not be measured by using an optical microscope. The laser beam was passed multiple times at laser power of 1000 and 1500 W to analyze the influence of the number of passes while keeping the same values of other parameters. When a better result was obtained with two passes of the laser beam at 1000 W power, two passes of 1500 W power laser beam were made at the same location shown in Fig. 2b. In this case, as observed through optical micrographs, the final borehole radius obtained was 6.28 mm. This implies that the radius can be increased with multiple passes though it will consume more power.

The preliminary inspection implied that little success was achieved for through-thickness penetration in free alignment or deep borehole creation. The single application of laser beam is not effective. Low penetration depth is observed confined to surface level. Though the depth may increase with multiple applications, the power consumption is higher. This leads to inefficiency in cost and productivity.

For boulder material, a 1500 W laser power provided better material removal. Hence, the same laser power value was employed in the construction marble sample. In this case, the aim was to vary the scanning speed and check its influence on material removal. In addition to other parameters as considered in the boulder sample case,

Fig. 4 Decoration marble material; **a** before, and **b** after laser interaction, circular, rectangular, and square shape of laser-processed areas; optical micrographs of laser-processed zones **c** circular, **d** rectangular, **e** square

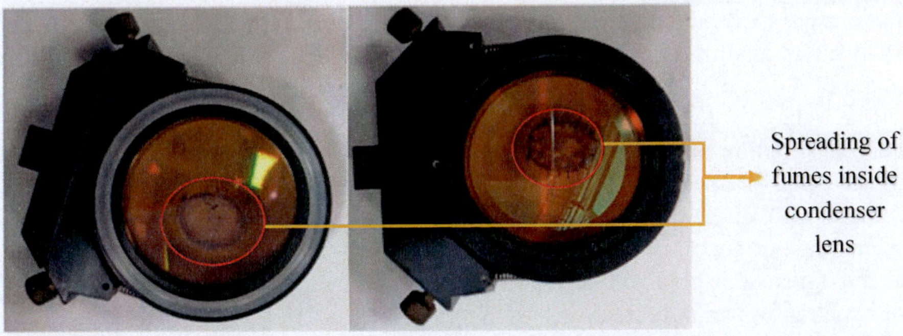

Fig. 5 Damage in condenser lens of 2.5 kW CO_2 laser machine due to sample reflections during scanning and nonlinear uneven surface profile of workpieces

Table 3 Process parameter for decoration marble interaction with laser

Operation attempts	Power (in watt)	Scanning speed (in mm/min)	Standoff distance (in mm)	Spot diameter (in mm)	Pressure (in bar)	Number of passes of laser beam
Circular scan shape	1500	600	1	0.5	20	1
Rectangular scan shape	1500	600	1	0.5	20	2
Square scan shape	1500	600	1	0.5	20	2

exposure time was also observed to check the time factor and heating effects. Free alignments were carried out for this sample at two different scanning speeds of 600 mm/min and 1200 mm/min at two different locations, as shown in Fig. 3b. It was found that with a lower scanning speed, the impact on material removal is higher, and the sample starts developing cracks, further breaking into two pieces after some time.

Further, borehole creation was attempted with a target hole diameter of 5 mm. Scanning speeds were also varied for this operation, where it was observed that, unlike boulder material, the borehole creation results are positive visually. Further on reducing the scanning speed to 600 mm/min, crack developments occurred, as shown in Fig. 3d.

Hence, with the construction marble material, the laser interaction yielded good results. With higher exposure time, variation in scanning speed and higher power values, the microcracks developed further, leading to a complete fracture of the sample. Better penetration ability was observed visually.

A decoration marble sample was selected to check the feasibility of employing an optical microscope for analyzing penetration depth. Deng et al. [9] highlighted the influence of laser scan shape on the drilling performance. They found the rectangular shape to be better in material removal. Hence, referring to the same, in the present work, the material removal capability of different laser-processed zone shapes was analyzed qualitatively, as depicted in Fig. 4. A laser power of 1500 W and a 600 mm/min scanning speed was used for laser processing all three shapes. These values were selected as they yielded better results in the case of boulder and construction marble samples. The optical microscope was employed to check the prospect of analyzing the drilling performance of LASER for laboratory-scale samples. However, the objective of using optical microscopy for depth measurement was not successful. It is useful in measuring the drilling diameter and scan shape dimensions obtained during laser interaction. However, measuring the depth of penetration was not effective.

The performance of LASER in drilling different samples presented here may depend on the hardness of the sample, mineral content, heterogeneity, thickness, and porosity of the sample. Further investigations are proposed to be carried out for characterizing the samples using Brinell hardness tester, X-ray diffraction (XRD), and scanning electron microscopy (SEM) to have a deeper understanding of the influence of laser power and material properties drilling depth.

In addition to the above results, using a 2.5 kW continuous mode CO_2 laser for ablating the discussed samples harmed the condenser lens of the machine. The reflections from the samples lead to unavoidable partial fuming of the lens, as shown in Fig. 4. Further scanning of samples with reflective nonlinear surface profiles can lead to fume leakage, which is fatal and hazardous. Apart from reflections, the fume formation and spreading can also be attributed to the nonlinear uneven surface profile of irregular-shaped materials like boulders and marbles. The same may happen for future targeted rock materials for further investigations. Hence, the proper profiling of the sample surface for scanning must be taken care of for laboratory-scale investigations.

The current work aimed to investigate the feasibility of laser material processing in removing non-metallic materials like boulder and marble using thermal energy generation to spall, melt, and vaporize the material. The oil and well drilling industries are primarily concerned about reducing specific energy consumption while increasing the rate and depth of penetration in fossil fields. These vital parameters are co-dependent on multiple factors from the aspect of material properties and process parameters. Material properties are anyhow independent factors and are difficult to control during interactions. However, if optimized, process parameters can be controlled as per material requirements and generate excellent possibilities for the usage of lasers in rock-drilling applications. The parametric analysis from available literature and the present work implies a trade-off situation for finding an optimal balance between laser power, exposure/irradiation time, and scanning speed. Hence, the trade-off must be dealt with through further investigations of material interaction and process parameters. A combinatorial approach for developing comprehensive geo-mechanical and physics-driven models must be employed. Such comprehensive models can effectively contribute to the economic viability of employing lasers in oil and gas well industries.

4 Conclusions

In conclusion, there are a few crucial observations enlisted as following.

- Multiple passes of laser are required in boulder material for producing a thorough drill. Pulse mode lasers are expected to provide better laser power and repetition control. Investigating the influence of using a laser beam in a continuous and pulse mode on the material removal shall provide better insight into this.
- During laser-rock interaction, different stages are observed as thermal heating, spallation, melting, and vaporization. The laser processing parameters need optimization concerning the temperature zones of these stages. Different materials may have different temperature ranges depending on their mechanical and thermal properties. Thermal analysis of different rock samples using a thermocouple or IR camera can provide information regarding the same.
- The optical phenomena such as scattering, radiation, and absorption depend on color contrast and define the total effective energy during the interaction. They are material dependent. Hence, material for laser interaction should be selected accordingly, considering the optical coefficient values.
- Optical microscopy fails to measure the penetration depth for laboratory-scale rock samples due to non-uniform beam distribution over irregular rock surface profiles.
- The irregular surface profile of rocks adversely affects the machine's condenser lenses, leading to unavoidable fume spread inside the lens. This can be taken care of by surface profiling of the samples and using regular-shaped samples with even surfaces for laboratory-scale investigations with continuous mode CO_2 lasers.

Acknowledgements This work is supported by IITG Technology Innovation and Development Foundation (IITGTI&DF), which has been set up at IIT Guwahati as a part of the National Mission on Interdisciplinary Cyber Physical Systems (NMICPS). IITGTI&DF is undertaking research, development and training activities on Technologies for Under Water Exploration with the financial assistance from Department of Science and Technology, India through grant number DST/NMICPS/TIH12/IITG/2020. Authors gratefully acknowledge the support provided for the present work. Laser facility used in this work was established with the support of Department of Science and Technology, India, under the first stage of FIST.

References

1. Graves RM, Batarseh S (2002) Application of high power laser technology to laser/rock destruction: Where Have We Been? Where Are We Now? AAPG Search Discovery. Article #90023©2002 AAPG Southwest Section Meeting. Ruidoso, New Mexico (2003), 213–224
2. Salavati S (2022) Applied high power laser in drilling oil and gas wells 15. https://doi.org/10.1117/12.2608976
3. Zediker MS (2014) High power fiber lasers in geothermal, oil and gas. Fiber Lasers XI Technol Syst Appl 8961:89610D. https://doi.org/10.1117/12.2042339

4. Graves RM, O'Brien DG (1998) StarWars laser technology applied to drilling and completing gas wells. In: Proceedings—SPE Annual Technical Conference and Exhibition, Delta, pp 761–770. https://doi.org/10.2118/49259-ms
5. Ahmadi M, Erfan MR, Torkamany MJ, Sabbaghzadeh J (2012) The effect of confining pressure on specific energy in Nd:YAG laser perforating of rock. Opt Laser Technol 44:57–62. https://doi.org/10.1016/j.optlastec.2011.05.017
6. Bharatish A, Kishore Kumar B, Rajath R, Narasimha Murthy HN (2019) Investigation of effect of CO_2 laser parameters on drilling characteristics of rocks encountered during mining. J King Saud Univ Eng Sci 31:395–401. https://doi.org/10.1016/j.jksues.2017.12.003
7. Bakhtbidar M, Abdideh M (2017) The effect of laser irradiation on rock permeability enhancement to improve oil recovery. In: 79th EAGE Conference on Exhibition, 2017. https://doi.org/10.3997/2214-4609.201700991
8. Li Q, Zhai Y, Huang Z, Chen K, Zhang W, Liang Y (2022) Research on crack cracking mechanism and damage evaluation method of granite under laser action. Opt Commun 506:127556. https://doi.org/10.1016/j.optcom.2021.127556
9. Deng R, Liu J, Kang M, Zhang W (2022) Simulation and experimental research of laser scanning breaking granite. Opt Commun 502:127403. https://doi.org/10.1016/j.optcom.2021.127403
10. Yan F, Gu Y, Wang Y, Wang C, Hu X, Peng H, Yao Z, Wang Z, Shen Y (2013) Study on the interaction mechanism between laser and rock during perforation. Opt Laser Technol 54:303–308. https://doi.org/10.1016/j.optlastec.2013.06.012
11. Schneider M, Berthe L, Fabbro R, Muller M (2008) Measurement of laser absorptivity for operating parameters characteristic of laser drilling regime. J Phys D Appl Phys 41. https://doi.org/10.1088/0022-3727/41/15/155502
12. Lin CH, Powell RA, Jiang L, Xiao H, Chen SJ, Tsai HL (2010) Real-time depth measurement for micro-holes drilled by lasers. Meas Sci. Technol 21. https://doi.org/10.1088/0957-0233/21/2/025307
13. O'Brien DG, Graves RM, O'Brien EA (1999) Laser-rock-fluid interaction: application of free-electron laser (FEL) in petroleum well drilling and completions. Free Laser Challenges II(3614):168–176. https://doi.org/10.1117/12.352664
14. Graves RM, Ionin AA, Klimachev YM, Mukhammedgalieva AF, O'Brien DG, Sinitsyn DV, Zvorykin VD (2000) Interaction of pulsed CO and CO_2 laser radiation with rocks typical of an oil field. High-Power Laser Ablation II(3885):159. https://doi.org/10.1117/12.376959
15. Gahan BC, Parker RA, Batarseh S, Figueroa H, Reed CB, Xu Z (2001) Laser drilling: determination of energy required to remove rock. In: Proceedings—SPE Annual Technical Conference and Exhibition, pp 1211–1221. https://doi.org/10.2118/71466-ms
16. Mahto V, Siddique H, Pathak AK (2012) Application of high power laser in oil and gas well drilling: an overview, 1–3, Proceedings of the National conference on Advances in Lasers and Spectroscopy (ALS-2012) 01-03 November, 2012, ISM Dhanbad, India
17. Martínez-Conde A, Krenke T, Frybort S, Müller U (2017) Review: comparative analysis of CO_2 laser and conventional sawing for cutting of lumber and wood-based materials. Wood Sci Technol 51:943–966. https://doi.org/10.1007/s00226-017-0914-9

Summary of Efforts in Phase Prediction of High Entropy Alloys Using Machine Learning

Swati Singh, Shrikrishna Nandkishor Joshi, and Saurav Goel

1 Introduction

Conventional alloys made from iron, aluminium, titanium, magnesium, and many others are prepared by mixing one or sometimes two principal elements with some secondary alloying elements in small proportions. Cantor alloys also recognized now as high-entropy alloys shows a unique characteristic that by mixing at least five metallic elements such that when the mixing percentage vary from 5 to 35% in an equimolar composition produces a single solid-solution phase which is crystalline in nature [1]. By virtue of this, HEAs offer high specific strength [2, 3], excellent stability at high temperatures, exceptional ductility [4], and high corrosion resistance and others [5]. Unlike conventional alloys, HEAs possess a trend of 'stronger being more ductile' [6, 7].

Cantor suggests that as many as ~ 108 varieties of HEAs can be developed using 44–64 elements from the chemical periodic table [8]. Many of these HEAs are yet to be synthesized in the lab and Yeh [9] suggests that "still a lot more treasure exist in non-equimolar HEAs". In developing an alloy, a considerable

S. Singh · S. N. Joshi (✉) · S. Goel
Department of Mechanical Engineering, Indian Institute of Technology Guwahati, Guwahati, Assam, India
e-mail: snj@iitg.ac.in

S. Singh
e-mail: sswati@iitg.ac.in

S. Goel
e-mail: goeLs@Lsbu.ac.uk

S. Goel
School of Engineering, London South Bank University, 103 Borough Road, London SE1 0AA, UK

Department of Mechanical Engineering, University of Petroleum and Energy Studies, Dehradun 248007, India

S. N. Joshi et al. (eds.), *Low Cost Manufacturing Technologies*,
https://doi.org/10.1007/978-981-19-8452-5_4

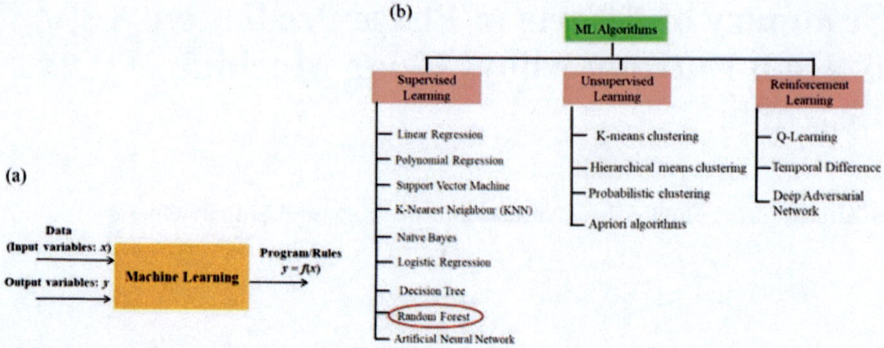

Fig. 1 **a** Illustrative architecture of ML, **b** classification of ML algorithms on the basis of supervised, unsupervised, and reinforcement learning [12–14]. Circle on Random forest is done to highlight that this method was used in this paper to present a case study

number of input parameters such as its composition, synthesis route, processing window, temperature, heating/cooling rate are required to obtain its crystallographic information and mechanical, electrical and functional properties of interest. Thus, relying on traditional laboratory experiments for novel material discovery can be very time-intensive.

Machine learning (ML) has emerged as a sophisticated and reliable technique in replacing repetitive laboratory experiments and computational simulations such as Density Functional Theory (DFT) and Molecular Dynamics (MD) [10]. DFT method predicts material properties using quantum mechanics and can at times be erroneous [11]. MD on the other hand analyse atoms by numerically solving Newton's equations of motion [12]. MD continues to suffer from the drawback on having reliable interatomic potential functions. Thus, ML provides a robust alternative tool based on the reliance of historical data and a mathematic way by pattern recognition technique. This in turn enhances our ability to extract salient features from within the data which are otherwise not readily visible even to an experienced researcher. A summary of various ML algorithms currently being used is shown in Fig. 1a which formed the core of this paper.

1.1 ML Algorithms

ML algorithms are broadly classified into three categories based on their type of learning, namely supervised, unsupervised, and reinforcement learning shown in Fig. 1b [13, 14].

1.1.1 Supervised Learning

As the name suggests, supervised ML algorithms are taught by providing a labelled dataset that includes input and output variables. The objective of supervised learning is to map input (x) and output (y), using a linear or nonlinear function, for instance, $y = f(x)$ [14].

1.1.2 Unsupervised Learning

In unsupervised ML algorithms, unlabelled dataset is used to train an algorithm, where inputs are not labelled with the correct outputs. The goal is to model the underlying structure or to discover the patterns in the data. The unlabelled data in the unsupervised learning is used to train algorithms for clustering and association problems.

1.1.3 Reinforcement Learning

Reinforcement learning is based on reward or punishment methods, where an agent learns to perceive and interpret its complex environment; it takes actions and learns through the trial-and-error method. It is devised to reward the desired behaviour by assigning a positive value to encourage the agent and punishing the undesirable behaviour by assigning negative values to penalize the agent. An agent either gets an award or a penalty based on the actions it performs, and the ultimate goal is to maximize the total reward. Over time, the agent learns to avoid the negative and seek the positive, thus learns the ideal behaviour to optimize its performance.

Current study focuses on a supervised ML algorithm, as the labelled data is employed for phase prediction of HEAs. A detailed description of all supervised ML algorithms has been discussed in Table 1.

1.2 Literature Review

Recently, there has been a surge in the number of publications on phase prediction of HEAs using various ML techniques. Islam et al. [15], Huang et al. [16] and Nassar et al. [17] employed neural networks in their study for phase prediction of HEAs and observed an average accuracy of 83%, 74.3%, and 90%, respectively, by using a relatively smaller number of datasets. These models considered five physical parameters namely mixing entropy (ΔSmix), valence electron concentration (VEC), atomic size difference (δ), mixing enthalpy (ΔHmix), and electronegativity difference ($\Delta \chi$). Choudhury et al. [18], used a random forest regression algorithm for the classification of different phases and crystal structures and obtained an accuracy of

91.66% for the classification of phases and 93.10% for the classification of different crystal structures, respectively.

To significantly improve the phase prediction accuracy, Risal et al. [19], compared Support Vector Machine (SVM), K-Nearest Neighbour (KNN), Random Forest (RF) classifier, and Multi-layer perceptron (MLP). KNN and RF Classifiers performed most effectively and obtained the test accuracy of 92.31% and 91.21%, respectively,

Table 1 A list of important supervised ML algorithms

Algorithms	Category	Description of the method	Advantages	Limitations
Linear regression	Regression	It correlates one independent variable and one dependent variable using a straight line	Easy to implement and understand, and it can extrapolate beyond a specific dataset	Assumes that dependent variables and independent variables are linearly related. Highly prone to noise and sensitive to outliers
Polynomial regression	Regression	Linear regression shows under fitting for nonlinear data, thus a new nth degree polynomial function is used to relate the independent and dependent variables	Capable of accommodating a wide range of functions	A correct polynomial function needs to be selected for better fitting of all the data points. It is more sensitive to outliers; even one or two outliers can significantly affect the outcome
Support vector machine	Regression and classification	It selects a decision boundary that best separates two different groups and predicts whether a new data falls into one category or the other	SVM performs well for high dimensional data as it provides various Kernels such as linear kernel, nonlinear kernel, polynomial kernel, Gaussian, radial basis function (RBF), and sigmoid kernel. It best suits binary classification problems	Performs poorly for overlapped classes. The selection of an appropriate Kernel and hyper-parameter is complex and problematic. A large dataset requires a long time for training

(continued)

Table 1 (continued)

Algorithms	Category	Description of the method	Advantages	Limitations
K-Nearest neighbour	Classification	Depending on the nearest neighbour, it classifies a new sample into one class out of several classes, separated depending on the type of data	Simple and easily accommodates itself when exposed to new data. Multiclass classification can be easily solved	It essentially requires feature scaling, performs poorly on imbalanced data, and cannot handle outliers and missing values
Naïve Bayes	Classification	It is based on 'Bayes theorem', assuming that no features are dependent and each feature is given the same weightage. It can classify data into several categories following the highest possibility	Simple and useful for vast datasets. It is insensitive to irrelevant features. Very fast, scalable, and effective for multiclass classification problems	Its application in the real world is limited because it assumes that all the features are independent and each feature makes an equal contribution
Logistic regression	Regression	An efficient method for binary and linear classification problems, it calculates probability from logistic regression equation to calculate the relationship between input variable and one or more output variables	It is simple, effective, and does not require feature scaling or hyper-parameter tuning	It performs poorly for nonlinear, irrelevant, and highly-correlated data

(continued)

Table 1 (continued)

Algorithms	Category	Description of the method	Advantages	Limitations
Decision tree	Regression and classification	Creates a tree-like structure	It is easy to visualize and interpret the results	It is data-sensitive, which means the outcome might change significantly on changing data slightly, and it is prone to overfitting
Random forest	Regression and classification	It takes votes from various decision trees for classification, and the average of all votes for a regression task	It can smoothly tackle the highly correlated features. It easily handles missing values, a large amount of data, or even imbalanced data and reduces the chances of overfitting, as the final decision depends on the decision of multiple trees	It is difficult to interpret its inner working; appears as a black box
Artificial neural network	Regression and classification	It is similar to human brain functioning and connects neurons to pass information	Performs well even with incomplete data. It works efficiently in recognizing patterns	Its functioning depends on the processing power. No specific rules are defined for selection of neurons or hidden layers. It is truly a black box algorithm

whilst SVM and MLP provided satisfactory performance with accuracies greater than 90%. Several similar studies have recently reported about the phase prediction and various mechanical properties which are summarized briefly in Table 2.

From the aforementioned literature, the random forest algorithm was observed to be the most prominently used algorithm in phase prediction studies, due to its high predictive performance. Therefore, the present study used random forest algorithm with the motive of correctly classifying each phase of HEAs for an imbalanced dataset. Scarce studies on the interpretation of the inner-working of an algorithm have been found in the literature. Thus, an additional attempt to decipher the black-box nature of the employed random forest algorithm has also been made using SHapely Additive exPlanation (SHAP) technique.

1.3 Data Source

A dataset of 1360 HEA samples was used to classify different phases of HEAs [30]. This dataset was observed to be highly imbalanced, as it contained 463 Intermetallic (IM), 441 BCC solid-solution (BCC_SS), 354 FCC solid-solution (FCC_SS), and

Table 2 Application of various ML algorithms in material science

Reference	Supervised ML method	Brief description	Success rate/Accuracy
Tancret et al. [20]	Gaussian processes (GP) + CALPHAD	Study on formation of single-phase solid-solution by considering the previously proposed empirical rules such as Hume-Rothery rules, basic thermodynamic concepts, and CALPHAD	63–80%
Islam et al. [15]	Neural network	Classification of solid-solution, intermetallic, and amorphous phase	83%
Huang et al. [16]	Comparison of k-nearest neighbour (KNN), support vector machine (SVM), and artificial neural network (ANN)	Solid-solution, intermetallic or mixed solid-solution and intermetallic phase formation study	SVM—64.3%, KNN—68.6%, ANN—74.3%
Choudhury et al. [18]	Random forest	Phase selection and crystal structure prediction study	91.66% for phase and 93.10% for crystal structure prediction
Li et al. [21]	Support vector machine (SVM)	Classification of BCC, FCC, and other phases not forming single phase solid-solution in CoCrFeMnNi HEA	96.55% training accuracy and 90.69% validation accuracy
Qi et al. [22]	Random forest	A phenomenological method to predict phases of high-entropy alloys using phase diagram data	> 80%
Zhou et al. [23]	Comparison of artificial neural network (ANN), convolutional neural network (CNN), and support vector machine (SVM)	Design rules for different phases (solid-solution (SS), amorphous (AM), and intermetallic (IM)) of high-entropy alloys	Accuracy of ANN model—98.9% for AM phase, 97.8% for SS phase, and 95.6% for IM phase

(continued)

Table 2 (continued)

Reference	Supervised ML method	Brief description	Success rate/Accuracy
Agrawal et al. [24]	ANFIS—Adaptive neuro fuzzy interface system	Classification of different phases of HEAs, using two approaches, one considering composition (composition-based model) as the inputs, whilst the other considering a set of six crucial parameters (parametric based model) in the formation of HEAs	84.21% for composition-based model, 80% for parametric-based model
Abdoon et al. [25]	Deep neural networks (DNNs)	Phase prediction of HEAs to design alloys	90%
Dai et al. [26]	Logistic regression	Phase prediction study on 407 data of high-entropy alloys using feature engineering	86% accurate with 9 descriptors
Kaufmann and Vecchio [27]	Random forest	High-throughput "ML-HEA" prediction of solid-solution forming-ability for HEAs by coupling thermodynamic and chemical features	(94% for binary) and (82.1% for ternary), when compared to the predictions from CALPHAD
Zhang et al. [28]	Support vector machine (SVM) model using four feature variables and Kernel principal component analysis (4 V-KPCA)	Phase prediction using relationship between phases and nine thermodynamics properties of high-entropy alloys	Testing accuracy—0.9743
Buranich et al. [29]	Linear regression (LR), Random forest (RF), and Gradient boosting regression (GBR)	Designing and screening of new HEAs for application in mechatronics industry	Highest accuracy (above 91%) for GBR
Risal et al. [19]	SVM, KNN, RF Classifiers, and ANN	Phase prediction of HEAs with an aim to significantly improve the phase prediction accuracy	KNN—92.31%, RF—91.21%, SVM and ANN—90%

(continued)

Table 2 (continued)

Reference	Supervised ML method	Brief description	Success rate/Accuracy
Nassar et al. [17]	Two neural networks: NN1 (considering only the compositional data), and NN2 (considering Hume-Rothery (HR) features along with the compositional data)	Phase prediction of HEAs	NN1—92%, NN2—90%
Machaka et al. [30]	Decision tree (DT) classifier, and Random forest (RF) classifier	Phase prediction of high-entropy alloys	DT—73%, RF—85%
Bhandari et al. [31]	Random forest	Yield strength prediction at the desired temperature for HEAs	93–97%, when compared with the experimental report for validation
Lee et. al. [32]	Regularized deep neural network (DNN) model	Identification of the key design parameters for enhancing the performance of phase prediction of HEAs. Furthermore, to overcome the problem of data shortage, a conditional generative adversarial network (GAN) was employed to generate more data	84.75%, 93.17% (when augmented by GAN)
Zeng et al. [33]	eXtreme Gradient Boosting (XGBoost) method	Phase selection rules for identification of single and mixed-phase	> 90%
Krishna et al. [34]	Performance comparison of logistic regression (LR), decision tree (DT), support vector machine (SVM), random forest (RF), gradient boosting classifier (GB), and artificial neural network (ANN)	Phase prediction in multiphase alloy system	LR—62.89%, SVM—83.02%, DT—77.99%, RF—82.39%, GB Classifier—81.13%, ANN—80.50%

102 mixed (FCC + BCC) phase. Five crucial features such as valence electron concentration (VEC), electronegativity difference ($\Delta\chi$), atomic size difference (δ), mixing enthalpy (dH_{mix}), and mixing entropy (ΔS_{mix}) calculated by Miedema's model were used as input parameters in the dataset. The ML modelling framework for the classification of four different phases of HEAs are demonstrated in Fig. 2, where data preprocessing is performed to detect outliers and missing values, and then feature scaling was performed to scale down the data into a finite range.

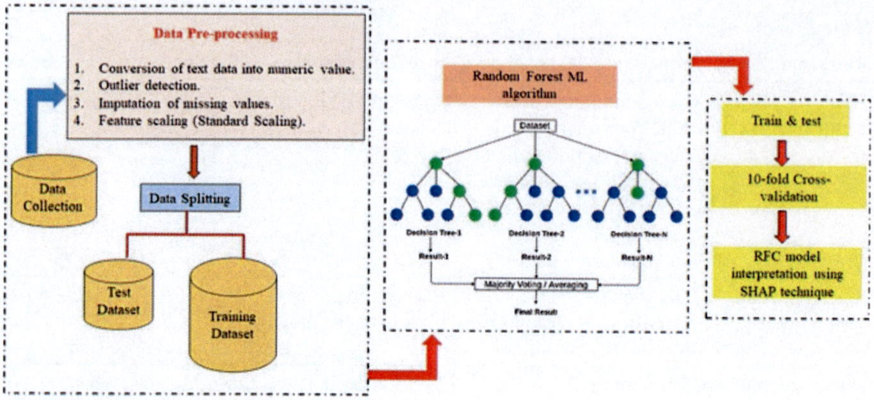

Fig. 2 ML modelling framework for phase classification of high-entropy alloys (HEAs)

The data was then divided into training and test sets (80:20) for the training (1088 HEA samples) and testing (272 HEA samples) purpose of the model. Random forest classifier (RFC) was employed using the scikit-learn library in Python, and further training and testing was performed to evaluate the performance of the RFC model.

1.4 RFC Model Performance

RF is an ensemble of various decision trees (tree-like structures) based on various subsets of the given dataset. For a classification task, it takes votes from various decision trees and makes a final prediction on the basis of majority votes. It is more accurate compared to a single decision tree algorithm, as a large number of trees improve its performance and makes the prediction more stable. The performance of RFC model has been evaluated using a confusion matrix and classification report. Confusion matrix provides the number of correctly predicted and incorrectly predicted classes for each phase [35]. Classification report on the other hand provides precision, recall, f1-score, average classification accuracy, and weighted average accuracy. Out of 272 HEA samples from the test data, 93 samples belong to class 0, i.e. IM phase, 88 samples belong to class 1, i.e. BCC phase, 71 samples belong to class 2, i.e. FCC phase, 20 samples belong to class 3, i.e. FCC + BCC mixed phase, as shown in classification report in Fig. 3a. The confusion matrix in Fig. 3b demonstrates the correctly and incorrectly classified phases, for example out of a total of 93 samples of IM phase, only 72 samples were correctly classified as IM phase, remaining 13 samples were misclassified as BCC phase, 5 samples were misclassified as FCC phase and 2 samples were misclassified as FCC + BCC phase. Similarly, the classification of each phase can be observed. Apart from an average classification accuracy of 86%, the precision, recall, and f1-score for all four phases were investigated in the classification report.

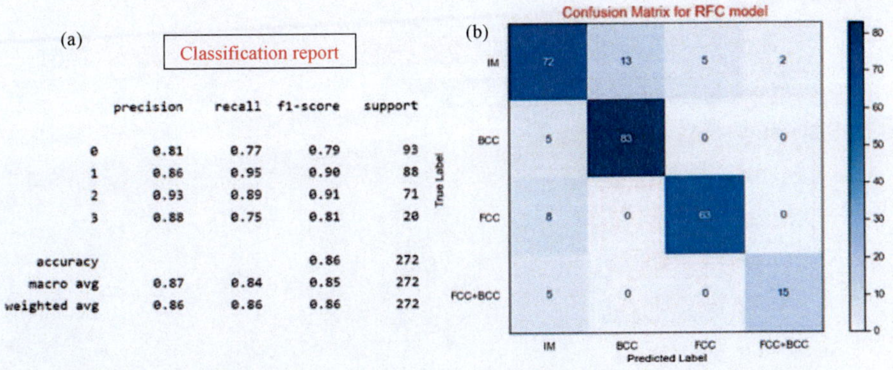

Fig. 3 **a** Confusion matrix, **b** classification report for RFC model in classification of IM, BCC, FCC, and FCC + BCC phase

Training accuracy, test accuracy, ROC_AUC_Score, and tenfold cross-validation ROC_AUC_Score were evaluated and shown in Table 3. The Receiver Operating Characteristic_Area under Curve (ROC_AUC) score ensures better performance of model in predicting different classes of HEAs. tenfold cross-validation was performed to avoid overfitting, where the complete dataset is divided into ten equal folds. Each time, the model was trained with 9 of those folds and the remaining onefold was used for the testing purpose, and the procedure was repeated ten times, by reserving a different tenth fold each time for testing the model. This measure ensured the effectiveness of the model and that the RFC model was not overfitting. The reported training accuracy of 90.9% and testing accuracy of 85.23%, suggested that the model was successfully classifying different phases of HEAs. The difference in the training and test accuracy indicated that the data used to train and test the RFC model was slightly different, which means that the RFC model is capable of predicting phases successfully, even for the unseen data points that have not been used in the present study.

Furthermore, the interpretation of RFC model was performed using SHAP technique, to understand which physical features were influential in governing phases of HEAs. SHapely Additive exPlanation (SHAP) technique has emerged as breakthrough in the field of ML for easy interpretation and explanation of a complex model's prediction; published by Lundberg and Lee in 2017 [36]. SHAP value aims to explain the prediction of an instance by considering the contribution of each feature

Table 3 Comparison of various ML algorithm performance

S. No.	Algorithm	Accuracy
1	Support vector machine	0.79
2	Decision tree classifier	0.81
3	Random forest classifier	0.86
4	XGBoost classifier	0.83

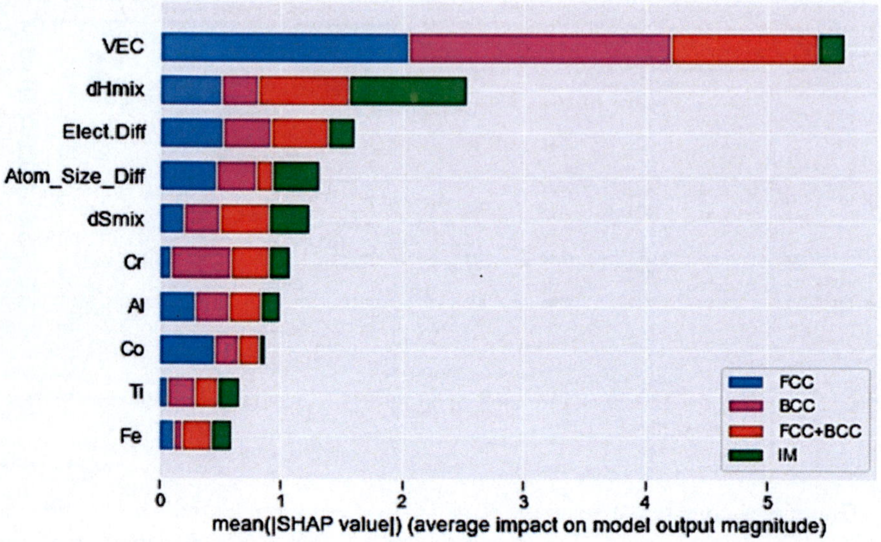

Fig. 4 Feature importance plot for RFC model using SHAP

Table 4 RFC model performance

S. No.	Evaluation	Model performance
1	Training accuracy	0.909672
2	Test accuracy	0.856088
3	ROC_AUC_Score	0.964626
4	Tenfold cross-validation ROC_AUC_Score	0.90311

in making a certain prediction at global as well as local levels. From this analysis, VEC was found to play most crucial role in determining FCC, BCC, and mixed FCC + BCC solid-solution phases whilst mixing enthalpy (dH_{mix}) was found important in determining the formation of solid-solution or an intermetallic phase, as shown in Fig. 4. Tables 4 and 5 show the RFC model performance and cross-validation score in each fold of tenfold cross-validation respectively.

2 Conclusions

The RFC model developed in this study had successfully and reliably predicted BCC, FCC, intermetallic, and FCC + BCC phases in high-entropy alloys. RFC model performance was evaluated using various evaluation metrics such as average classification accuracy, precision, recall, f1-score, ROC_AUC score, and tenfold cross validation ROC_AUC score.

Table 5 Cross-validation score in each fold of tenfold cross-validation

No. of folds	Cross-validation score
Fold 1	0.88633
Fold 2	0.93082
Fold 3	0.78310
Fold 4	0.92372
Fold 5	0.94252
Fold 6	0.93863
Fold 7	0.83451
Fold 8	0.89151
Fold 9	0.96391
Fold 10	0.93750

Mean value of tenfold cross-validation score: 0.9031
Standard deviation: 0.0533

The RFC model showed that the two input parameters, namely valence electron concentration (VEC) and mixing enthalpies were most influential in determining the resulting phase of a given HEA composition. VEC contributed the most in predicting the crystal structure of solid solution phases (BCC, FCC, and FCC + BCC) whilst mixing enthalpy (ΔH_{mix}) played important role in determining formation of solid-solution or intermetallic. Thus, this present study leveraged the reliability of applying ML techniques in material science without any requirement of performing expensive experiments.

Acknowledgements Swati Singh greatly acknowledge the scholarship provided by the Ministry of Education, Government of India. Saurav Goel greatly acknowledge the support provided by the Royal Academy of Engineering via Grants No. IAPP18-19\295 and TSP1332.

References

1. Katiyar N, Goel G, Goel S (2021) Emergence of machine learning in the development of high entropy alloy and their prospects in advanced engineering applications. Emergent Mater 4(6):1635–1648. https://doi.org/10.1007/s42247-021-00249-8
2. Senkova ON, Senkova SV, Woodward C, Miracle DB (2013) Low-density, refractory multi-principal element alloys of the Cr–Nb–Ti–V–Zr system: Microstructure and phase analysis. Acta Mater 61(5):1545–1557. https://doi.org/10.1016/j.actamat.2012.11.032
3. Stepanov N, Shaysultanov D, Salishchev G, Tikhonovsky M (2015) Structure and mechanical properties of a light-weight AlNbTiV high entropy alloy. Mater Lett 142:153–155. https://doi.org/10.1016/j.matlet.2014.11.162
4. Deng Y, Tasan C, Pradeep K, Springer H, Kostka A, Raabe D (2015) Design of a twinning-induced plasticity high entropy alloy. Acta Mater 94:124–133. https://doi.org/10.1016/j.actamat.2015.04.014
5. Zhang Y, Li R (2020) New advances in high-entropy alloys. Entropy 22(10):1158. https://doi.org/10.3390/e22101158

6. Tsai M, Yeh J (2014) High-entropy alloys: a critical review. Mater Res Lett 2(3):107–123. https://doi.org/10.1080/21663831.2014.912690
7. Youssef K, Zaddach A, Niu C, Irving D, Koch C (2014) A novel low-density, high-hardness, high-entropy alloy with close-packed single-phase nanocrystalline structures. Mater Res Lett 3(2):95–99. https://doi.org/10.1080/21663831.2014.985855
8. Cantor B (2021) Multicomponent high-entropy Cantor alloys. Prog Mater Sci 120:100754. https://doi.org/10.1016/j.pmatsci.2020.100754
9. Yeh J (2006) Recent progress in high-entropy alloys. Annales De Chimie Science Des Matériaux 31(6):633–648. https://doi.org/10.3166/acsm.31.633-648
10. Kremer K, Grest G (1990) Molecular dynamics (MD) simulations for polymers. J Phys Condens Matter 2(S):SA295-SA298. https://doi.org/10.1088/0953-8984/2/s/045
11. Neugebauer J, Hickel T (2013) Density functional theory in materials science. Wiley Interdiscip Rev: Comput Mol Sci 3(5):438–448. https://doi.org/10.1002/wcms.1125
12. Goel S, Knaggs M, Goel G, Zhou X, Upadhyaya H, Thakur V et al (2020) Horizons of modern molecular dynamics simulation in digitalized solid freeform fabrication with advanced materials. Mater Today Chem 18:100356. https://doi.org/10.1016/j.mtchem.2020.100356
13. Osisanwo FY, Akinsola JE, Awodele O, Hinmikaiye JO, Olakanmi O, Akinjobi J (2017) Supervised machine learning algorithms: classification and comparison. Int J Comput Trends Technol 48(3):128–138. https://doi.org/10.14445/22312803/ijctt-v48p126
14. Nasteski V (2017) An overview of the supervised machine learning methods. HORIZONS B 4:51–62. https://doi.org/10.20544/horizons.b.04.1.17.p05
15. Islam N, Huang W, Zhuang H (2018) Machine learning for phase selection in multi-principal element alloys. Comput Mater Sci 150:230–235. https://doi.org/10.1016/j.commatsci.2018.04.003
16. Huang W, Martin P, Zhuang H (2019) Machine-learning phase prediction of high-entropy alloys. Acta Mater 169:225–236. https://doi.org/10.1016/j.actamat.2019.03.012
17. Nassar A, Mullis A (2021) Rapid screening of high-entropy alloys using neural networks and constituent elements. Comput Mater Sci 199:110755. https://doi.org/10.1016/j.commatsci.2021.110755
18. Choudhury A, Konnur T, Chattopadhyay P, Pal S (2019) Structure prediction of multi-principal element alloys using ensemble learning. Eng Comput 37(3):1003–1022. https://doi.org/10.1108/ec-04-2019-0151
19. Risal S, Zhu W, Guillen P, Sun L (2021) Improving phase prediction accuracy for high entropy alloys with machine learning. Comput Mater Sci 192:110389. https://doi.org/10.1016/j.commatsci.2021.110389
20. Tancret F, Toda-Caraballo I, Menou E, Rivera Díaz-Del-Castillo P (2017) Designing high entropy alloys employing thermodynamics and Gaussian process statistical analysis. Mater Des 115:486–497. https://doi.org/10.1016/j.matdes.2016.11.049
21. Li Y, Guo W (2019) Machine-learning model for predicting phase formations of high-entropy alloys. Phys Rev Mater 3(9). https://doi.org/10.1103/physrevmaterials.3.095005
22. Qi J, Cheung A, Poon S (2019) High entropy alloys mined from binary phase diagrams. Sci Rep 9(1). https://doi.org/10.1038/s41598-019-50015-4
23. Zhou X, Zhu J, Wu Y, Yang X, Lookman T, Wu H (2022) Machine learning assisted design of FeCoNiCrMn high-entropy alloys with ultra-low hydrogen diffusion coefficients. Acta Mater 224:117535. https://doi.org/10.1016/j.actamat.2021.117535
24. Agrawal A, Choudhary A (2016) Perspective: materials informatics and big data: realization of the "fourth paradigm" of science in materials science. APL Mater 4(5):053208. https://doi.org/10.1063/1.4946894
25. Abdoon Al-Shibaany Z, Alkhafaji N, Al-Obaidi Y, Atiyah A (2020) Deep learning-based phase prediction of high-entropy alloys. IOP Conf Ser: Mater Sci Eng 987(1):012025. https://doi.org/10.1088/1757-899x/987/1/012025
26. Dai D, Xu T, Wei X, Ding G, Xu Y, Zhang J, Zhang H (2020) Using machine learning and feature engineering to characterize limited material datasets of high-entropy alloys. Comput Mater Sci 175:109618. https://doi.org/10.1016/j.commatsci.2020.109618

27. Kaufmann K, Vecchio K (2020) Searching for high entropy alloys: a machine learning approach. Acta Mater 198:178–222. https://doi.org/10.1016/j.actamat.2020.07.065
28. Zhang L, Chen H, Tao X, Cai H, Liu J, Ouyang Y et al (2020) Machine learning reveals the importance of the formation enthalpy and atom-size difference in forming phases of high entropy alloys. Mater Des 193:108835. https://doi.org/10.1016/j.matdes.2020.108835
29. Buranich V, Rogoz V, Postolnyi B, Pogrebnjak A (2020) Predicting the properties of the refractory high-entropy alloys for additive manufacturing based fabrication and mechatronic applications. In: IEEE international conference on "nanomaterials: applications & properties" (NAP-2020) symposium on additive manufacturing and applications (SAMA-2020) Sumy, Ukraine, 9–13 Nov 2020
30. Machaka R (2021) Machine learning-based prediction of phases in high-entropy alloys. Comput Mater Sci 188:110244. https://doi.org/10.1016/j.commatsci.2020.110244
31. Bhandari U, Rafi M, Zhang C, Yang S (2021) Yield strength prediction of high-entropy alloys using machine learning. Mater Today Commun 26:101871. https://doi.org/10.1016/j.mtcomm.2020.101871
32. Lee S, Byeon S, Kim H, Jin H, Lee S (2021) Deep learning-based phase prediction of high-entropy alloys: optimization, generation, and explanation. Mater Des 197, 109260. https://doi.org/10.1016/j.matdes.2020.109260
33. Zeng Y, Man M, Bai K, Zhang Y (2021) Revealing high-fidelity phase selection rules for high entropy alloys: a combined CALPHAD and machine learning study. Mater Des 202, 109532. https://doi.org/10.1016/j.matdes.2021.109532
34. Krishna Y, Jaiswal U, Rahul R (2021) Machine learning approach to predict new multiphase high entropy alloys. Scripta Mater 197:113804. https://doi.org/10.1016/j.scriptamat.2021.113804
35. Markoulidakis I, Rallis I, Georgoulas I, Kopsiaftis G, Doulamis A, Doulamis N (2021) Multiclass confusion matrix reduction method and its application on net promoter score classification problem. Technologies 9(4):81. https://doi.org/10.3390/technologies9040081
36. Lewis F, Butler A, Gilbert L (2010) A unified approach to model selection using the likelihood ratio test. Methods Ecol Evol 2(2):155–162. https://doi.org/10.1111/j.2041-210x.2010.00063

Conceptual Design of Extrusion Systems for Cement Paste 3D Printing

Shubham Maurya, Biranchi Panda, Uday Shanker Dixit, Arun Ch. Borsaikia, and Biswajeet Barman

1 Introduction

Three-dimensional (3D) printing of cement-based materials is an emerging digital technology, enabling the fabrication of structures layer-by-layer without any formwork [1–5]. The production parameters in 3D printing can be grouped into two main categories: the printer and the mixed design parameters. While all these parameters can be altered, it is essential to develop an ecosystem with optimized and improved coordination of the 3D printer and the material. A typical 3D printer uses robotic or gantry systems, mounted with a print head that extrudes the material layer by layer directly from 3D model. There are different extrusion mechanisms studied in the literature such as ram extruder, augur screw extruder, cavity pump (progressive cavity) extruder, a combination of augur screw with progressive cavity extruder, and pneumatic driven extruder [6, 7]. The ram extruders use a driving mechanism to drive the plunger to achieve reciprocating movement. The linear movement of the ram displaces the material and results in forwarding extrusion as shown in Fig. 1a. The working principle of the pneumatic-based extruder is similar except the fact that compressed air is used to move the piston as shown in Fig. 1b. The geometry of augur screw-based extruder is very simple. It consists of an augur or Archimedes screw, and its rotational motion displaces the material in forward direction toward the nozzle (see Fig. 1c). Figure 1d shows an extrusion system based on principle of a progressive cavity pump. It is made up of helical rotor and double helix stator. The stator and rotor are designed in such a way that it forms cavities in between

S. Maurya (✉) · B. Panda · U. S. Dixit · B. Barman
Department of Mechanical Engineering, Indian Institute of Technology Guwahati, Guwahati, Assam 781039, India
e-mail: shubhamaurya128@iitg.ac.in

A. Ch. Borsaikia
Department of Civil Engineering, Indian Institute of Technology Guwahati, Guwahati, Assam 781039, India

S. N. Joshi et al. (eds.), *Low Cost Manufacturing Technologies*,
https://doi.org/10.1007/978-981-19-8452-5_5

and material flows in the cavity toward the nozzle by means of the rotating rotor [8]. The flow rate is directly proportional to the rotation speed of rotor. To improve the material intake capacity and efficiency, it can be further modified by combining with augur screw as shown in Fig. 1e.

The extruder is the soul of any 3D printer (material extrusion type) and used for extruding the material through the nozzle orifice for depositing the material layer-by-layer manner. In this paper, a systematic approach is used for converting extruder design requirements into a final concept selection using the well-known Pugh's method of concept selection [9–11]. The ultimate goal of conceptual design is to select the most desirable concept for smooth and continuous extrusion of the cement paste.

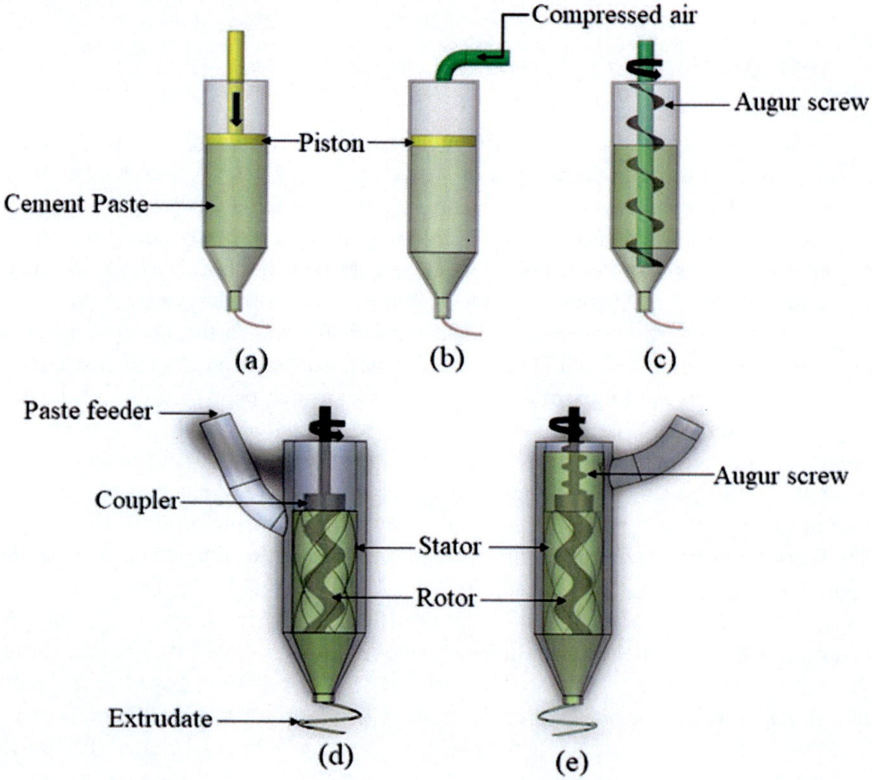

Fig. 1 Type of extrusion mechanism: **a** ram-based, **b** pneumatic-based, **c** augur screw-based, **d** cavity pump-based extruder, **e** combination of augur screw with cavity screw

Table 1 Mix proportion of 3D printable cement paste

MIX ID	Binder	HPMC (%binder)	Water/binder
M1	1.0	–	0.32
M2	1.0	0.1	0.30

2 Materials

In this study, ordinary Portland cement (OPC 43 grade) manufactured by Dalmia Cement (Bharat) Ltd. was used as binder, while hydroxy-propyl methyl cellulose (HPMC) was used (only in M2 mix) as viscosity modifying agent with specific gravity of 3.12 and 1.26, respectively. Tap water was used for all the mixtures. The binder and water were mixed in three steps with a Borosil Hand Blender. This includes slow speed mixing for 1 min, medium speed mixing for 2 min, scraping for 30 s, and finally high-speed mixing for 1 min. Many trial mixes were designed to optimize the mix proportion with respect to "extrudability" and "buildability" criteria. Table 1 shows the final mix proportion used for 3D printing testing.

3 Methodology

The design of extruder prototype includes several steps like the identification of a need, a literature survey of an existing design, the development of the new idea for the identified need, the decision of the best idea, and then validation [7]. The methodology followed in this paper to design and develop the cement paste extruder is shown in Fig. 2.

3.1 Identification of Need

Before starting design of the extruder, the real needs were established. Digital fabrication in the construction industry is growing rapidly due to the advantage like design freedom, low production cost, and reduced manufacturing time. In this context, the authors understand the need for development of cement paste extruder for educational training purposes. This helps in understanding the working principle of extrusion-based 3D printer.

3.2 Consideration of Alternative Designs

There are different types of paste extruder available with different extruding mechanism; however, "syringe-based" and "augur screw-based extruder" design concepts

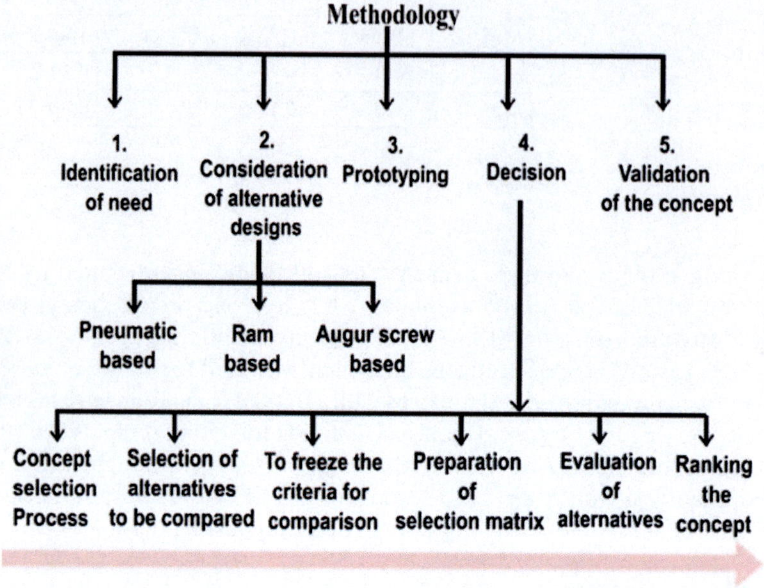

Fig. 2 Methodology used to meet the objective

were considered based on comprehensive literature review. Figure 3 shows the schematic of body structure and working mechanism of the two extruder concepts. The ram-based extruder uses a nut and bolt (threaded rod) mechanism to convert the rotation of motor to linear motion. There are three main components: (1) motor mounting and mechanism to transfer the rotation into linear motion (2) syringe holder which allows the syringe to be replaced with another syringe (3) syringe with nozzle.

3.3 Prototyping and Testing

This section describes the development of prototype of final product and testing of fabricated model to analyze the working of cement paste extruder. Section 3.3.1 describes the development of prototype, and Sect. 3.3.2 describes the testing of the prototype.

3.3.1 Prototyping

Prototype is the pre-launch version of the complete product used for extensive testing and trial of the product. While designing the concept, authors used fused deposition modeling (FDM) technique to fabricate the components to avoid any design errors.

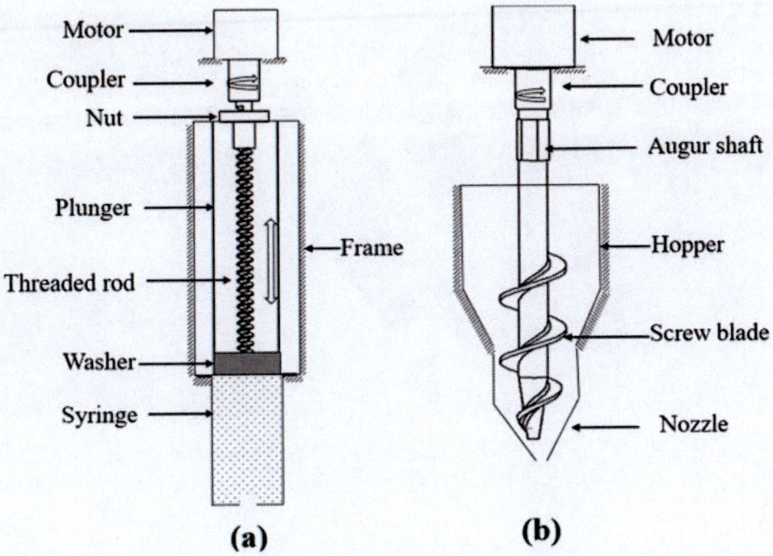

Fig. 3 Working mechanism of **a** ram-based, **b** augur screw-based extruder

Figures 4 and 5 show the computer-aided design (CAD) and physical model of ram and screw-based extrusion concepts, respectively.

The following components were printed using FDM process to fabricate the extruder:

(1) Holding frame (for both the motor and the syringe)
(2) Plunger
(3) Plunger holder
(4) Augur screw

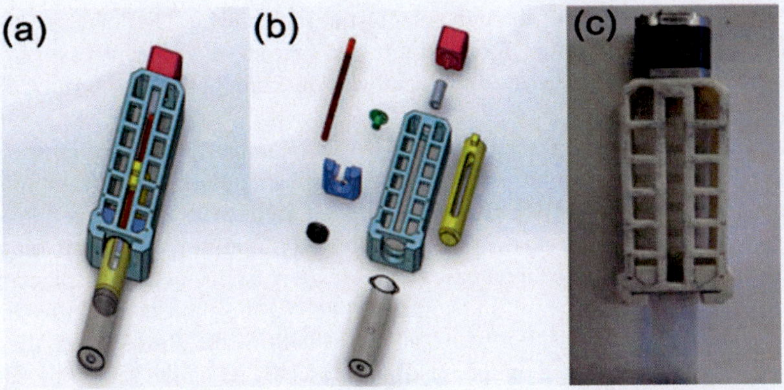

Fig. 4 CAD model and physical model of augur screw extrusion system; **a** CAD assembly, **b** exploded view of assembly, **c** physical model assembly

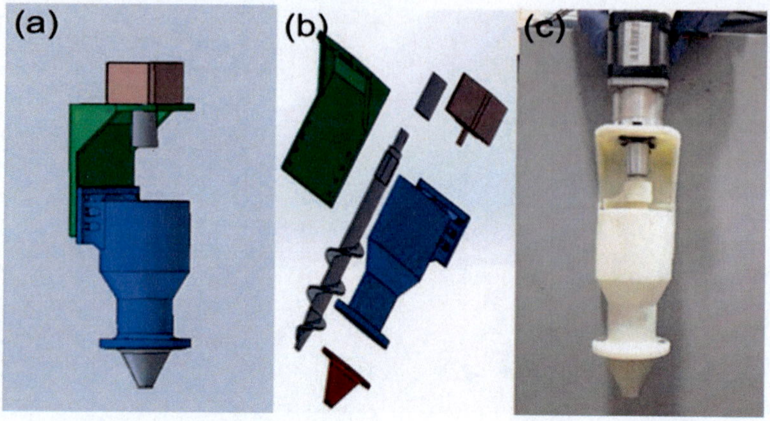

Fig. 5 CAD model and physical model of augur screw-based extrusion system; **a** CAD assembly, **b** exploded view of assembly, **c** physical model assembly

(5) Hopper (100 ml)
(6) Nozzle with different exit diameter.
(7) Holding frame (augur screw-based design for hopper and motor).

The remaining parts were obtained from local vendors, which include 150 mm ACME 4-start lead screw with lead of 8 mm and pitch of 2 mm with copper nut, a 100 ml Braun Omniana Syringe (latex-free), and a NEMA 17 geared stepper motor. The complete assembly of the components is shown in Figs. 4c and 5c.

3.3.2 Cement Paste Extrusion Testing

The selection of motor torque capacity is an essential task to drive the plunger in a linear direction to push the material out of the nozzle. Therfore, an experiment was carried out with the help of the universal testing machine (UTM) machine to know the maximum and minimum force required to extrude the cement paste from the extruder. Two syringes were used with and without cement paste material (M1 mix) as shown in Fig. 6.

In order to analyze the force required to push the material from the nozzle, linear displacement of the piston at a constant ram displacement rate of 20 mm/min was applied. It was found that the maximum force required to move the piston inside the syringe without cement paste is 39 N, while the maximum force required to push the plunger in a syringe with cement paste was 377.76 N against a total displacement of 62 mm (Fig. 7). After 62 mm, the cement paste becomes too compact by forming a dead zone [12] which cannot be extruded. In both of the above cases, the motor torque was calculated based on the maximum force by using the following equation [13]:

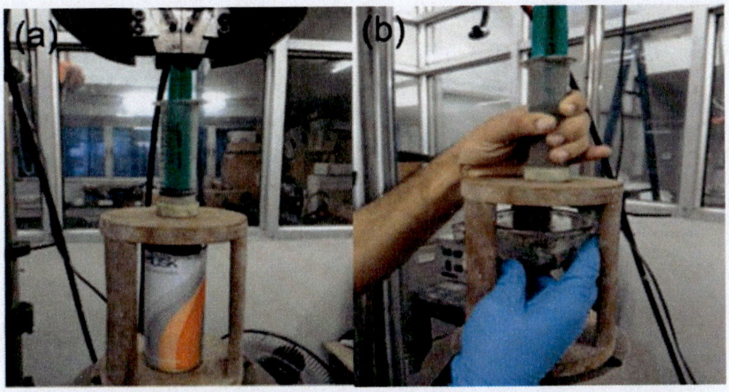

Fig. 6 Experiment for extrusion force estimation **a** without and **b** with paste

$$T_r = \frac{F d_m (L + \mu \pi d_m \sec \phi)}{2(\pi d_m - \mu L \sec \phi)} \quad (1)$$

where d_m is mean diameter of screw (= 7.0 mm), L is lead of screw (= 8 mm, for 4-start thread used here), μ is coefficient of friction (= 0.15), and ϕ is half the thread angle (= 14.5° for Acme thread used here).

Based on the maximum torque of 59 kg.cm, NEMA17 geared motor was selected for the extruders. This is planetary geared stepper motor with 1:19 gear ratio at rated torque of 5.88 N-m at 15 revolutions per minute (rpm).

Experiments were carried out to analyze the extrudability, buildability, and dead zone formation of cement paste in syringe corresponding to different nozzle size [12]. The dead zone is the zone in syringe where the materials to be extruded, comes to rest and resists the flow of material. The dead zone length must be as short as possible for optimal extrusion. It means that the mixture must be extrudable via a given nozzle without the need for excessive pressure, as the excessive pressure

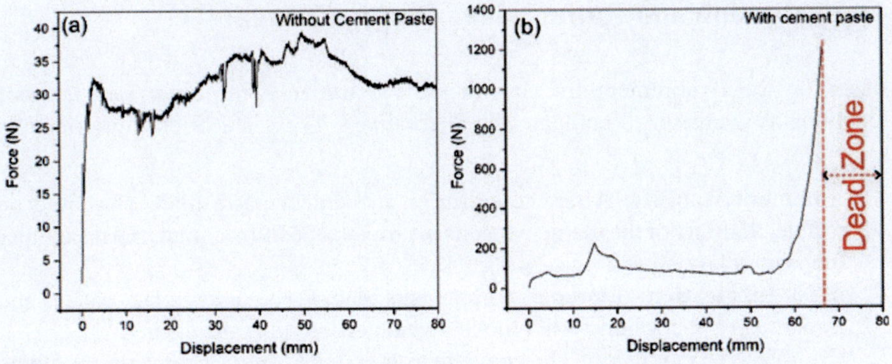

Fig. 7 Force versus displacement plot: **a** without cement paste, **b** with cement paste

affects the properties of the printed substance as well as the cement paste. If the dead zone is longer, it results in difficulty in extrusion, wastage of material, and also print quality. Thus, it is desirable for printable mixtures to demonstrate shorter dead zone lengths and longer plug flow zones. In Fig. 7b, the pressure for extrusion is suddenly increased in dead zone region; it is because of effect of consolidation and water filtration under the hydrodynamic conditions. Nair et al. [14] defined a geometric ratio given by

$$\psi = \frac{D - d_{\text{entry}}}{d_{\text{entry}}}\left(1 + \sqrt{\frac{\text{curved surface area of nozzle}}{A_{\text{exit}}}}\right) \quad (2)$$

where D is syringe diameter (= 30 mm), L is nozzle length ($L = D$), d_{entry} is the entry diameter of the nozzle (= 10 mm for both nozzles), d_{exit} is the exit diameter of the nozzle (= 4 mm for nozzle N10-4 and 5 mm for nozzle N10-5), and A_{exit} is the exit cross-sectional area. In line with Eq. 2 used by Nair et al. [14], nozzle N10-4 and N10-5 with a geometric ratio (ψ) of 16.52 and 14.0, respectively, for syringe were considered. The minimum dead zone was observed with nozzle with lower value of geometric ratio. The comparison was made between the mix M1 and Mix M2 for the dead zone length, and it was found that mix M2 will have lower dead zone length in both the nozzles. It is because of HPMC, which stops the water filtration. As per findings of Nair et al., the dead zone length is lower for the nozzle with low geometric ratio, which was also confirmed here.

The extrudability of ram-based extruder was confirmed by the extruder prototype. The motor was controlled by an Arduino microcontroller and stepper motor driver by fixing the motor speed to 3 rpm, which results in the linear displacement of the plunger with speed of 24 mm/min and flow rate of 5 mm^3/s as shown in Fig. 8a. For augur screw-based extruder, continuous filament extrusion was observed around 50 rpm motor speed as shown in Fig. 8b (Fig. 9).

4 Evaluation and Rating of the Design Concepts

Based on the requirement for cement paste extruder, some criteria in term of functionality were set to compare the alternatives. These are briefly described as follows:

1. **Filament stability**: After the extrusion of cement paste filament should be stable. Stability of the filament results in proper deposition of material on another deposited layer.
2. **Fast fabrication**: Extruder components should be assembled easily and mounted on the robotic arm quickly, which ensure the time saving.
3. **Bubble-free extrusion**: The cement paste extruded from the nozzle should be compact with no air bubble with it. This ensures the continuous extrusion of material without break in printing.

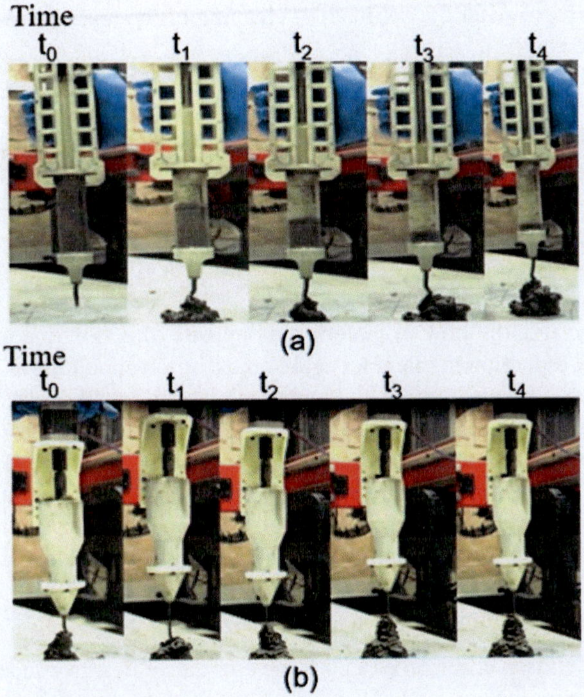

Fig. 8 Extrudability testing using **a** syringe, **b** screw-based extruder concept

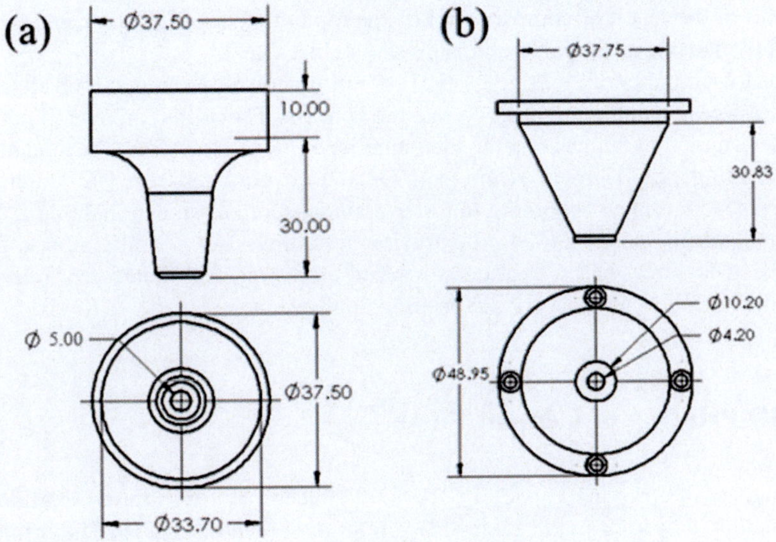

Fig. 9 Nozzle design used in **a** syringe, **b** hopper

4. **Ease to clean**: After the application cleaning of the nozzle and other extruders should be easy.
5. **Ease to feed the material**: Printing material feeding should be easy to ensure the continuous printing.
6. **Constant flow**: The flow rate and diameter of extruded material from the nozzle end should be constant.
7. **Size of extruder**: The size of the extrusion system must be in a way that it should be installed on a gantry-type robotic system.
8. **Minimum contact**: The extruder must be in less contact with cement paste which ensures the less dirty parts.
9. **Backflow**: It is the flow of material in opposite direction due to back pressure caused by reduction in diameter at the nozzle tip. It should be as less as possible in extruder because it will stop the extrusion of material and then printing.
10. **Dead Zone**: It is the region near the nozzle entry where the material in the syringe comes to rest (also called static zone) when the material is forced under pressure because of the sudden change in cross-section area at the starting of nozzle. The dead zone length is completely dependent on the nozzle, mixture microstructural packing, and beneficial rheology. For the cement-based materials, the dead zone length can be controlled by controlling microstructural packing and beneficial rheology, which is desired in this application.

Each of the chosen criterion of a possible alternative design concept is compared with the corresponding criterion of the currently used datum, and the result is recorded in the decision matrix as (✓) if more favorable, (×) if less favorable, and (=) if the same. The decision on whether a concept is better than the datum is based on the analysis of the result of comparison, i.e., the total number of (✓), (×), and (=). The decision matrix was prepared and is presented in Table 2.

The total number of "✓" ratings and "×" ratings corresponding to each alternative concept were computed in order to rate the concept. Thereafter, an overall net score was determined by subtracting the total number of "×" ratings from the total number of "✓" ratings, i.e., [Sum (✓) Sum (×)]. Once the overall net score is calculated, the concepts were ranked according to the overall net score as shown in Table 2. Based on the obtained score, ram-based extrusion system was found to be more effective compared to screw-based extruder for paste extrusion, and therefore, "ram-based extruder" was further used for 3D printing application.

5 3D Printing of Cement Paste

To evaluate the printing capability, the final concept, i.e., ram-based extruder was mounted on a custom-made gantry 3D printer as shown in Fig. 10. The printer has 3 degree of freedom, i.e., the extruder can travel in X, Y, and Z direction. Printing speed is crucial when using paste materials rather than thermoplastic filament. Most of the non-Newtonian paste-like materials have varied curing times, which means

Table 2 Decision matrix

Selection criteria	1st alternatives	2nd alternatives	Datum
Stability to extrudate	=	=	Stable
Quick fabrication	=	=	Less time
Backflow	✓	×	No
Constant flow rate	✓	×	Yes
Bubble-free extrusion	×	✓	Maximum
Cleaning after use	✓	×	Easy
Feed	×	✓	Easy
Extruder size	=	=	less
Contact	✓	×	Minimum
Dead zone	×	✓	Minimum/less
Sum (✓)	4	3	
Sum (×)	3	4	
Sum (=)	3	3	
Overall score	+1(✓)	− 1(×)	
Rank	1st	2nd	

the printer only has a limited amount of time to produce a component before the substance begins to cure or harden, making processing difficult or impossible [15]. The M2 mix was used for printing as it contains HPMC which thickens the paste and thereby enables smooth extrusion and stable print quality. In order to verify the performance of the developed extruder, square shape designs were printed with various infill patterns such as (a) wedge shape and (d) grid shape with a thickness of 5 mm. The printed designs were shown in Fig. 11.

6 Conclusions

The conceptual design of cement paste extruder for 3D printing, based on the well-known Pugh's method, was very effective for comparing extruder design concepts. In light of the approach taken, the syringe-based extruder was found to be more suitable for cement paste extrusion compared to screw-based design as confirmed by printing trails. In addition to extruder design, material rheology is a key parameter followed by nozzle shape and size. Future work can be focused on numerical flow

Fig. 10 3D cement paste printer with developed ram-based extruder

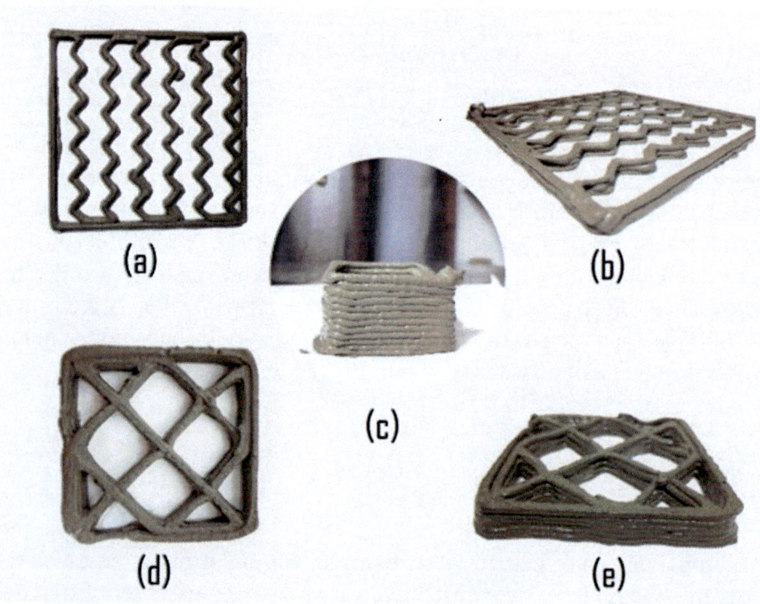

Fig. 11 3D printed shapes; **a** square (10 cm × 10 cm × 1 cm) filled with wedge shape-top view, **b** square (10 cm × 10 cm × 1 cm) filled with wedge shape-isometric view, **c** square 5 cm × 5 cm × 4 cm with no infill, **d** square (10 cm × 10 cm × 1 cm) with grid shape infill-top view, **e** square (10 cm × 10 cm × 1 cm) with grid shape infill-isometric view

analysis for better understanding of material flow in the extruder by variable nozzle shape and size.

Acknowledgements This work is a part of the project entitled "Design and development of an intelligent extrusion device for 3D printing of concrete structures." The authors would like to acknowledge the financial support from the Department of Science and Technology (DST) India through grant number DST: TPN:58083.

References

1. ISO/DIS 17296-1(en), Additive manufacturing—general principles—Part 1: Terminology. https://www.iso.org/obp/ui/#iso:std:iso:17296:-1:dis:ed-1:v1:en. Accessed 2 June 2022
2. Tofail SA, Koumoulos EP, Bandyopadhyay A, Bose S, O'Donoghue L, Charitidis C (2018) Additive manufacturing: scientific and technological challenges, market uptake and opportunities. Mater Today 21(1):22–37
3. Holzmann P, Breitenecker RJ, Soomro AA, Schwarz EJ (2017) User entrepreneur business models in 3D printing. J Manuf Technol Manag
4. Tay YWD, Panda B, Paul SC, Noor Mohamed NA, Tan MJ, Leong KF (2017) 3D printing trends in building and construction industry: a review. Virtual Phys Prototyp 12(3):261–276
5. Dey D, Srinivas D, Panda B, Suraneni P, Sitharam TG (2022) Use of industrial waste materials for 3D printing of sustainable concrete: a review. J Clean Prod 130749
6. Guo C-F, Zhang M, Bhandari B (2019) A comparative study between syringe-based and screw-based 3D food printers by computational simulation. Comput Electron Agric 162:397–404. https://doi.org/10.1016/j.compag.2019.04.032
7. El Mesbahi J, Buj-Corral I, El Mesbahi A (2020) Use of the QFD method to redesign a new extrusion system for a printing machine for ceramics. Int J Adv Manuf Technol 111(1):227–242
8. Gravesen J (2008) The geometry of the Moineau pump. Comput Aided Geom Des 25(9):792–800. https://doi.org/10.1016/j.cagd.2008.06.012
9. El Mesbahi J, Buj-Corral I, El Mesbahi A (2021) Design of an innovative new extrusion system for a printing machine for ceramics. Int J Adv Manuf Technol 117(1):591–603
10. Pugh S (1991) Total design: integrated methods for successful product engineering. Addison-Wesley
11. Frey DD, Herder PM, Wijnia Y, Subrahmanian E, Katsikopoulos K, Clausing DP (2007) An evaluation of the Pugh controlled convergence method. Int Des Eng Tech Conf Comput Inf Eng Conf 48043:193–203
12. Perrot A, Rangeard D, Nerella VN, Mechtcherine V (2018) Extrusion of cement-based materials—an overview. RILEM Tech Lett 3:91–97
13. Budynas RG, Nisbett JK (2011) Shigley's mechanical engineering design, vol 9. McGraw-hill, New York
14. Nair SAO, Panda S, Santhanam M, Sant G, Neithalath N (2020) A critical examination of the influence of material characteristics and extruder geometry on 3D printing of cementitious binders. Cem Concr Compos 112:103671. https://doi.org/10.1016/j.cemconcomp.2020.103671
15. Amza C, Zapciu A, Popescu D (2017) Paste extruder—hardware add-on for desktop 3D printers. Technologies 5(3):50. https://doi.org/10.3390/technologies5030050

Friction Stir Spot Welding of Honeycomb Core Sandwich Structure

A. Kumar, R. Ganesh Narayanan, and N. Muthu

1 Introduction

Sandwich honeycomb structures are extensively used in the aerospace and automotive industries due to their excellent bending stiffness and strength [1–4]. A sandwich is a composite made up of two faces, or sheets, separated by and joined to a low stiff and low dense core. Sandwich cores and skins may be connected with adhesives and mechanical fasteners. Although the adhesive bonding gives the benefit of a continuous joint, it is time-consuming and costly due to the extensive surface preparation and curing time. Further, toxic chemicals in the adhesive can negatively impact human health and the environment [5]. Mechanical deformation of rivet has been used to weld sandwich sheets [6]. A hole in the rivet tail develops due to the buckling of the rivet tail during the procedure. The residual compressive force is relaxed as a result of the hole creation. Fatigue causes joints to loosen. Additionally, in SPR, the upper sheet is also deformed non-uniformly. It is also possible to attach the thick core to the sandwich panels with mechanical fasteners [7]. However, predrilled holes for mechanical fastenings cause stress concentrations in the region through the hole, and the consumable fasteners also add weight to the structures. As a result, a new way of joining the honeycomb core sandwich sheets must be developed.

FSSW process is used to join the honeycomb core and sheets. FSSW is a solid-state joining method where a rotating tool plunges and retracts into overlapping core and sheets to form the joint [8–10]. The procedure is carried out in three stages:

A. Kumar (✉) · R. Ganesh Narayanan · N. Muthu
Department of Mechanical Engineering, Indian Institute of Technology Guwahati, Guwahati, Assam 781039, India
e-mail: avneesh.kumar@iitg.ac.in

R. Ganesh Narayanan
e-mail: ganu@iitg.ac.in

N. Muthu
e-mail: nelsonm@iitg.ac.in

S. N. Joshi et al. (eds.), *Low Cost Manufacturing Technologies*,
https://doi.org/10.1007/978-981-19-8452-5_6

plunging, stirring, and retracting [11]. The plunging stage is divided into two stages: In the first stage, the tool rotation is restricted, and the tool movement is allowed to plunge up to a certain plunge depth. In the second stage, the rotation of the tool starts, and then plunging is done up to a particular plunge depth. The FSSW provides several advantages over traditional joining methods, including less distortion, longer fatigue life, no consumables required, and environmental friendliness. It is considered that various difficulties such as cavity creation and weight addition may be avoided by utilizing this procedure for welding. Even though the current analysis includes several FSSW parameters, this is a preliminary effort to investigate the effect of rotational speed on plunge force, torque, and lap-shear test performance during FSSW of sandwich sheets.

2 The Experimental Process

This section discusses the methods used for the welding process, specimen preparation, load and torque measurement during welding, and mechanical testing of the welded joint.

2.1 Specimen Preparations

The specimen in this study is made from a 0.95 mm thick AA5052-H32 alloy sheet and a 6 mm thick honeycomb core made of AA3003 alloy. Honeycomb core is sandwiched between two AA5052-H32 sheets to create the sandwich sheet. No adhesive is used at any point of contact to prevent the adhesive from affecting the joint behavior. The metallic sheets are sheared into suitable dimensions using a leg shearing machine and a hand shearing cutter for the honeycomb core. Figure 1 shows the dimensions of the lap-shear specimen schematically.

2.2 Welding Joint Preparations

The FSSW joints are produced using a vertical type three-axis friction stir welding machine (FSW). Acetone is used to thoroughly clean all surfaces of metallic sheets to eliminate dirt and oil. All of the relevant specimen pieces are overlappingly arranged on the bed of the machine. There is no movement in any direction since the specimen is completely clamped. As shown in Table 1, three separate parameter sets are used for the joining. Only the rotational speed varies between 233, 342, and 462 revolutions per minute. Other parameters such as plunge speed, dwell time, and plunge depth are maintained constant. FSSW joints are made using a straight cylindrical pin having a

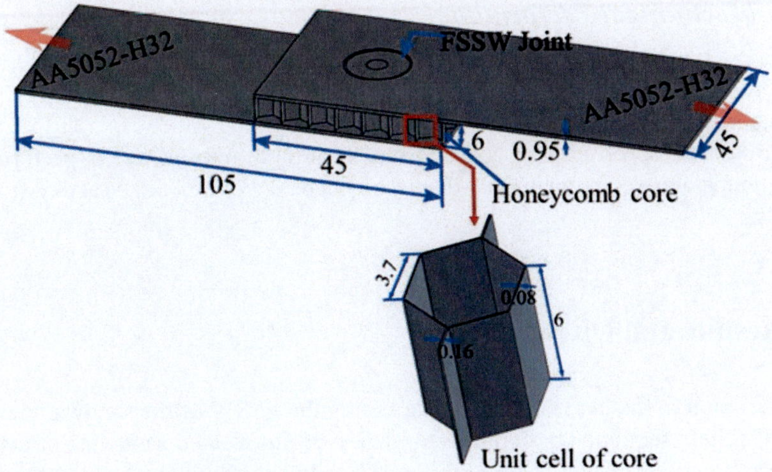

Fig. 1 Lap-shear specimen (all dimensions are in mm)

Table 1 Parameters used during welding

S. No.	Rotational speed of the tool (rpm)	Plunge depth of the tool (without rotation) (mm)	Plunge depth of the tool (when rotation starts) (mm)	Tool plunge speed (mm/min)	Dwell time (s)
1	233	6.2	1.5	2	5
2	342	6.2	1.5	2	5
3	462	6.2	1.5	2	5

flat shoulder shape. The pin diameter, shoulder diameter, and pin length are 5 mm, 16 mm, and 1.5 mm, respectively.

2.3 Load Response Measurement

The load cell mounted to the machine measures the load and torque responses. During the procedure, four separate indices are monitored continuously against time. The variables to consider are rotational speed, axial tool movement, torque, and axial load. The variation in all indexes is displayed against time to show the effect of the process parameters.

2.4 Mechanical Performance

The produced joint is tested using a lap-shear test on a 100 kN capacity electro-mechanically operated Universal Testing Machine (UTM) (model: INSTRON-8801). The crosshead speed is held constant at 1 mm/min, and the test is performed at room temperature. Sandwich load–displacement behavior is obtained at three rotating speeds.

3 Results and Discussions

The variation in the load and the torque during the FSSW of the sandwich is investigated in this section. The joint performance of the welded sandwich structure is compared and correlated at three different rotational speeds by the lap-shear test. Lastly, the fracture behavior is studied on the fractured surfaces of the lap-shear tested sandwich structures.

3.1 Load and Torque Response

The load and torque responses with time during welding are plotted in Fig. 2. These graphs are plotted during plunging with the tool rotation. The plunging or axial load decreases with the increase in the rotational speed as the flow stress decreases at increased temperatures. In comparison, the maximum welding torque increases with the increase in the rotational speed. The maximum load of 489 N is reached at 233 rpm, while the maximum torque of 233 N-mm is obtained at 462 rpm. There is a fluctuation in both plunging load and torque during welding; it occurs because of the breaking of the honeycomb core into tiny particles as the honeycomb core has a significantly lower wall thickness of 0.08 mm.

3.2 Lap-Shear Performance

Joint strength is represented in the FSSW in terms of failure load since the actual area of load-bearing is unknown. The load–extension behavior of the honeycomb core sandwich structure in the lap-shear test at three rotational speeds is shown in Fig. 3. The maximum fracture load of 1338 N is obtained at 462 rpm, whereas the lowest fracture load of 350 N occurs at 233 rpm. The fracture load and failure displacement are increasing with rotational speed, and at all rotational speeds, the displacement during the failure is approximately in proportion to the highest load. This is associated

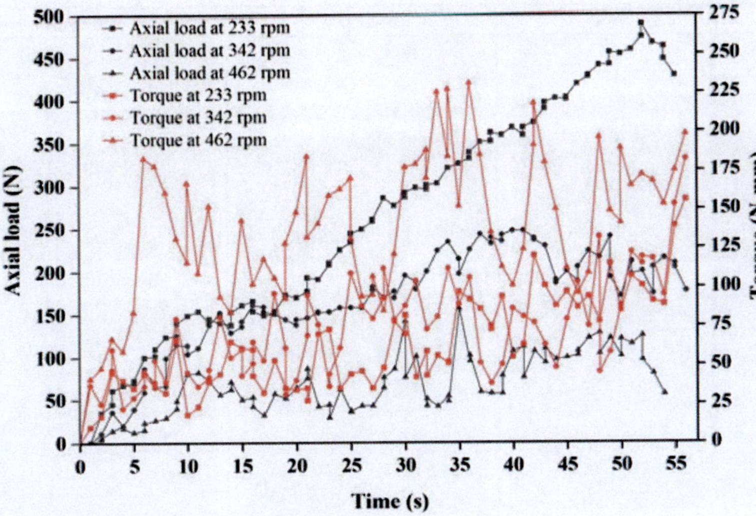

Fig. 2 Variation of the axial load and the torque during the FSSW

with the elongation of the welded part. FSSW joints with higher shear strength can be highly ductile, whereas poorly welded joints have lower ductility [12].

Figure 4 shows the failure modes after lap-shear tests. Nugget pull-out and shear failures are both types of failure encountered. A shear fracture mode is observed at a rotational speed of 462 rpm; however, nugget pull-out failure occurs at 233

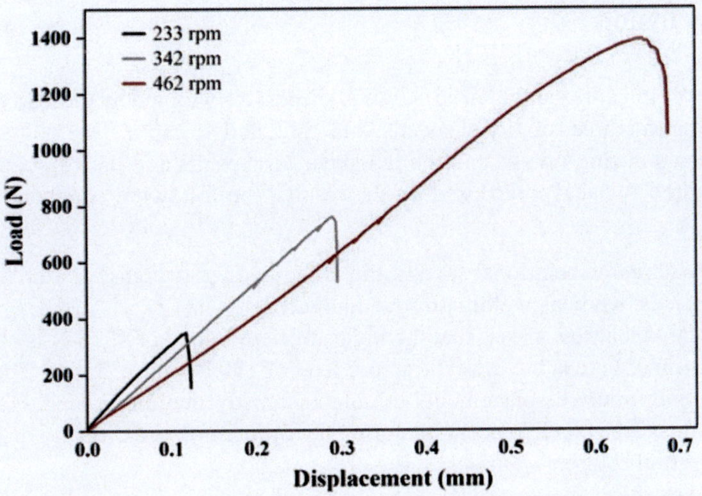

Fig. 3 Load evolution during lap-shear test of the FSSW welded sandwich structure

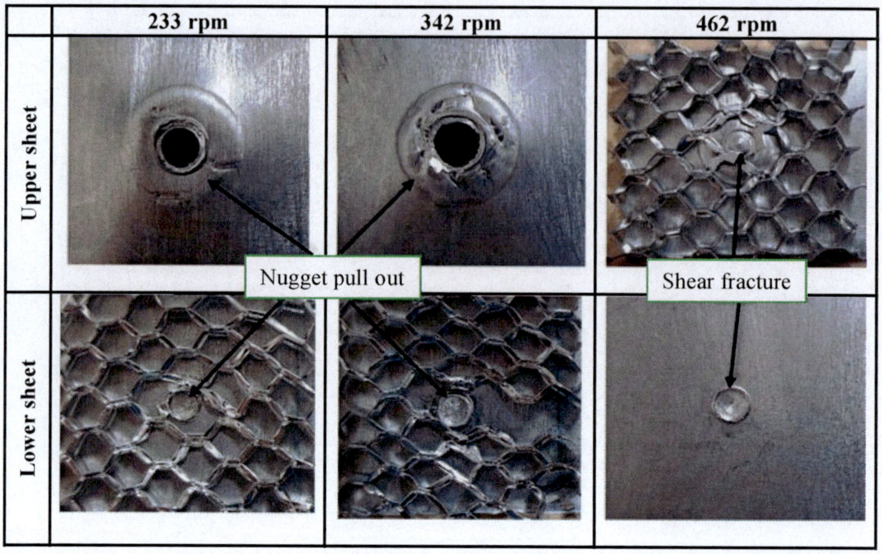

Fig. 4 Failure modes during lap-shear test

and 342 rpm. All plates and core are separated at lower rotational speeds (233 and 342 rpm), while the core is attached to the top sheet at the highest rotational speed, 462 rpm.

4 Conclusion

The honeycomb core sandwich structure is successfully joined by the FSSW process considering the three rotational speeds, 233, 342, and 462 rpm. The load and torque requirements during FSSW and the lap-shear test results and their fracture pattern are compared for analysis. Based on the results, the following conclusions can be drawn:

(a) With increased rotational speeds, the plunging load decreases as the flow stress decreases, whereas welding torque increases.
(b) The lap-shear test shows that the highest fracture load of 1338 N is obtained at 462 rpm, whereas the lowest fracture load of 350 N is at 233 rpm. The fracture load and failure displacement both increase with rotational speed. For all rotational speeds, the displacement during the failure is approximately proportional to the highest load.
(c) The lap-shear analysis revealed nugget pull-outs and shear failures. The shear fracture mode is observed at a rotational speed of 462 rpm, and nugget pull-out failure is observed at 233 and 342 rpm.

Acknowledgements The work is supported by the Science and Engineering Research Board, India, under the scheme 'Early Career Research Award', No: ECR/2018/001638.

References

1. Hara D, Özgen GO (2016) Investigation of weight reduction of automotive body structures with the use of sandwich materials. Transp Res Procedia 14:1013–1020. https://doi.org/10.1016/j.trpro.2016.05.081
2. Gibson LJ, Ashby MF (1997) Cellular solids: structure and properties. Cambridge Solid State Science Series
3. Hohe JO, Becker W, A re®ned analysis of the effective elasticity tensor for general cellular sandwich cores. [Online]. Available: www.elsevier.com/locate/ijsolstr
4. Liu Y, Liu W, Gao W (2021) Out-of-plane shear property analysis of Nomex honeycomb sandwich structure. J Reinf Plast Compos 40(3–4):165–175. https://doi.org/10.1177/0731684420943285
5. Barnes TA, Pashby IR (2000) Joining techniques for aluminum spaceframes used in automobiles. Part I—solid and liquid phase welding. J Mater Process Technol 99(1):62–71. https://doi.org/10.1016/S0924-0136(99)00367-2
6. Pickin CG, Young K, Tuersley I (2007) Joining of lightweight sandwich sheets to aluminium using self-pierce riveting. Mater Des 28(8):2361–2365. https://doi.org/10.1016/j.matdes.2006.08.003
7. De Matteis G, Landolfo R (1999) Mechanical fasteners for cladding sandwich panels: interpretative models for shear behaviour. Thin-Walled Struct 35(1):61–79. https://doi.org/10.1016/S0263-8231(99)00017-8
8. Rana PK, Narayanan RG, Kailas SV (2018) Effect of rotational speed on friction stir spot welding of AA5052-H32/HDPE/AA5052-H32 sandwich sheets. J Mater Process Technol 252:511–523
9. Gerlich A, Su P, North TH (2005) Tool penetration during friction stir spot welding of Al and Mg alloys. J Mater Sci 40(24):6473–6481. https://doi.org/10.1007/s10853-005-1568-9
10. Rana PK, Narayanan RG, Kailas SV (2021) Assessing the dwell time effect during friction stir spot welding of aluminum polyethylene multilayer sheets by experiments and numerical simulations. Int J Adv Manuf Technol 114(7–8):1953–1973. https://doi.org/10.1007/s00170-021-06910-0
11. Piccini JM, Svoboda HG (2015) Effect of the tool penetration depth in friction stir spot welding (FSSW) of dissimilar aluminum alloys. Procedia Mater Sci 8:868–877. https://doi.org/10.1016/j.mspro.2015.04.147
12. Mitlin D, Radmilovic V, Pan T, Chen J, Feng Z, Santella ML (2006) Structure-properties relations in spot friction welded (also known as friction stir spot welded) 6111 aluminum. Mater Sci Eng A 441(1–2):79–96. https://doi.org/10.1016/j.msea.2006.06.126

Low-Cost La(III)-Bentonite@Chitosan and La(III)-Bentonite@Polysulfone Composite Beads for the Removal of Dyes and Phosphate from Water Bodies

Moucham Borpatra Gohain, Diksha Yadav, Sachin Karki, Kongkona Gogoi, and Pravin G. Ingole

1 Introduction

Phosphate, dyes, and metal ions are some of the most toxic pollutants in water bodies due to their acute toxicity, carcinogenicity, and non-biodegradable nature. Nowadays, water scarcity has become a worldwide problem due to increasing world population, rising high living standards, and the need for clean water has pushed for immediate solutions [1]. Moreover, in the whole world, water scarcity is being observed as a present and future risk to human activity [2, 3]. According to the United Nations of Environment Programme (UNEP), 80% of the water in global is untreated and contains almost everything from human waste to industrial toxic discharge [4]. The water serves for drinking and other purposes; thus, the contaminated/polluted water has a direct impact on our habitat, and it is also dangerous for aquatic life. Pathogens (human and animal waste), organic matter, chemical pollution, and salinity are the major pollutants in water, and many of these kinds are severely dangerous to us [5, 6]. This waste generally contains toxic heavy metal ions, phosphate, and dyes present in it, and these ions contaminate the water for a long time and promote water pollution. With increasing clean water requirements and to meet its high demand, affordable water treatment is necessary. There are several choices for cleaning water and using polymeric composite beads for separating the impurities (phosphate and dyes) is an interesting option.

M. B. Gohain · D. Yadav · S. Karki · P. G. Ingole (✉)
Chemical Engineering Group, Engineering Sciences and Technology Division, CSIR-North East Institute of Science and Technology, Jorhat, Assam 785006, India
e-mail: pingole@neist.res.in

Academy of Scientific and Innovative Research (AcSIR), Ghaziabad, Uttar Pradesh 201002, India

K. Gogoi
Analytical Chemistry Group, Material Sciences and Technology Division, CSIR-North East Institute of Science and Technology, Jorhat, Assam 785006, India

S. N. Joshi et al. (eds.), *Low Cost Manufacturing Technologies*,
https://doi.org/10.1007/978-981-19-8452-5_7

Polysulfone (PSf) comes under the class of thermoplastics that have excellent thermal, chemical, and mechanical stability which makes it a strong candidate for the preparation of solvent-impregnated beads by the phase inversion method [7–11]. Polysulfone is easily dissolved in N,N-dimethylformamide (DMF) with immersed stirring under heating conditions. There are several reported beads based on polysulfone (PSf) and chitosan types of polymer and biopolymer, respectively. Chitosan is the material which is largely useful for the removal of heavy metal ions, dyes, phosphate, etc., from industrial wastewater. The use of chitosan as a bio-adsorbent also provides value addition to the water purification system [12]. Over the traditional methods, the chitosan-based water treatment methods are highly cost-effective and safe for the removal of phosphorus using chitosan/Ca-organically modified montmorillonite beads [13]. Due to the various advantages of chitosan as an environment-friendly, economical bio-sorbent, we used it in the current study for the removal of phosphate and dye molecules. As we know the removal of dyes from the wastewater is a great challenge and there are various methods or technologies/processes for it like biological [14], chemical [15], photocatalytic [16], ozone treatment [17], membrane [18, 19], and adsorption [16, 20]. Compared with the several reported methods clay-based adsorbents like montmorillonite (MMT), palygorskite, and bentonite have attracted more because of their low cost and availability in a wide range [21–23].

Here we have used bentonite as it is easily available and its nonmetallic mineral has a 2:1 type crystal structure composed of two tetrahedrons sandwiched by a layer of Al-oxide. It is having good cation exchange and adsorption capacity; due to these advantages, we used it for the removal of phosphate and dye molecules along with lanthanum doping. In this work, two types of composite beads have been prepared and applied for the sorption of dyes and phosphate under analogous experimental conditions. Considering the advantages mentioned above of bentonite, chitosan and polysulfone we applied it to make the composite beads with modification in internal microstructure and surface morphology by incorporating an additive, namely La-Bnt. The developed La-bentonite@polysulfone beads were characterized with various physiochemical characterization techniques like thermogravimetric analysis (TGA), Fourier transforms infrared spectroscopy (FTIR), and field emission scanning electron microscopy (FE-SEM). Additionally, Chitosan gel beads have also been prepared and tested and data has been reported in this paper.

2 Experimental Methodology

2.1 Materials

The chemical reagents used in the experiments are PSf (Polysulfone; M_w ~ 30,000) purchased from M/s Sigma Aldrich Chemical Company, USA, DMF solvent, chitosan nanopowder, lanthanum nitrate, and potassium dihydrogen phosphate (KH_2PO_4) purchased from Merck. The methyl violet (MV) dye was brought from

TCI Co. Ltd. NaOH, HCl, bentonite were purchased from SRL. The deionized water (DI) was obtained from the millipore available in our laboratory at CSIR-NEIST. Most of the chemicals were analytical grade and used directly.

2.2 Preparation of La-Bentonite (La-Bnt)

The same method used by Zhang et al. is implemented for the preparation of La-Bnt [24]. First, the bentonite was calcined by a muffle furnace for 2 h and impregnated into a 5% HCl solution to activate it. Then further washed with deionised water until pH gets neutral and later dried under vacuum. Furthermore, to make it La-Bnt, 0.5 g treated bentonite was added to an aqueous solution of lanthanum nitrate (0.3 wt%). Later the solution pH is adjusted to ~ 10 with NaOH solution (2 M). Then after a standstill for 12 h, the formed blending was washed with deionized water until neutral pH and calcinated at 500 °C. The obtained sample was noted as La-Bnt.

2.3 Preparation of La(III)-Bentonite@Polysulfone Beads

With La-Bnt nanomaterial, polysulfone beads were prepared. An 18% polysulfone (PSf) with 0.1 wt% of La-Bnt nanomaterials in DMF was prepared by continuously stirring on a magnetic stirrer at 600 rpm and 70 °C temperature. After that, the solution was kept for a while to get cool down so that if any moisture was present, it will get away and no air bubbles will be left in the polymer solution. The polymer solution was then filled into the syringe and dropped into a coagulation bath (water bath) dropwise. This causes the polysulfone to solidify as a result of the phase inversion method and occurs the formation of La(III)-bentonite@polysulfone composite beads. The beads were kept in the deionized water for 1 h to remove the remaining solvent and then dried for 48 h at room temperature. Figure 1 represents the fabrication process of La(III)-bentonite@polysulfone composite beads.

2.4 Preparation of La(III)-Bentonite@Chitosan Beads

Chitosan beads were prepared by incorporating La-Bnt nanoparticles, and the fabrication process used for the preparation was somewhat different in comparison with polysulfone-based beads. Figure 2 represents a schematic diagram of preparing La(III)-bentonite@chitosan beads. A chitosan solution (in acetic acid) is mixed with La-Bnt nanopowder until the solution becomes homogeneous. The solution consists of 1% v/v of acetic acid mixed with 3 wt% of chitosan incorporated with 0.1 wt% of La-Bnt nanopowder. The prepared solution was stirred for 10 h on a magnetic stirrer and afterward filled into a 10 ml syringe for further processing. The solution was

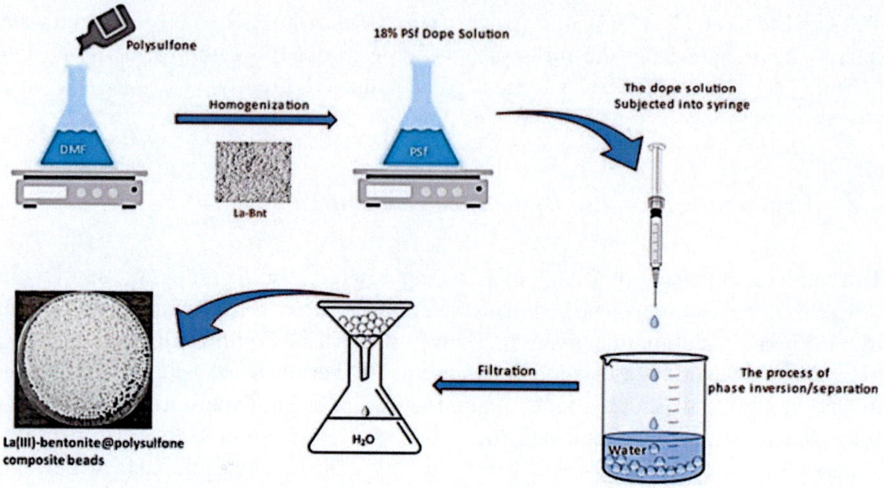

Fig. 1 Preparation of the La(III)-bentonite@polysulfone composite beads

added dropwise into a 0.5 M NaOH solution with the help of a syringe where the phase inversion process instantly occurred and chitosan beads were formed. After the formation of chitosan beads, they were kept in the NaOH solution for 24 h. After 24 h the beads were cleaned with DI water until the pH become neutral. Finally, the beads were air-dried until their weight become constant [25].

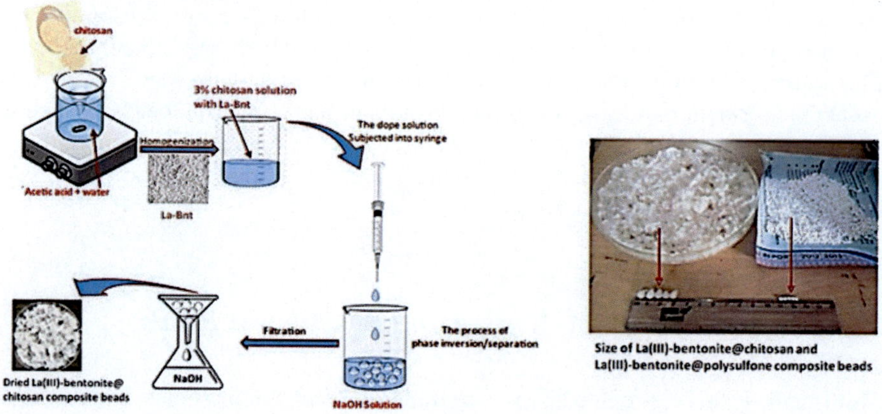

Fig. 2 Preparation method of La(III)-bentonite@chitosan beads

Table 1 Experimental results of La(III)-bentonite@chitosan and La(III)-bentonite@polysulfone composite beads for the removal of methyl violet dye

S. No.	Sample name	Feed concentration	Feed	Average removal efficiency %
1	La(III)-bentonite@polysulfone	30 ppm	Methyl violet	99.4
2	La(III)-bentonite@chitosan	30 ppm	Methyl violet	79.2

2.5 *Performance Study*

An adsorption study for 30 ppm methyl violet (MV) dye and 20 ppm phosphate solution (by dissolving KH_2PO_4 in deionized water) has been done. The above-mentioned concentration was added to the phosphate solution and dye solution for 1 h at room temperature at 540 rpm. After every 10 min, the mixture solution was filtered using filter paper and the collected samples were taken for UV analysis. Each experiment was repeated 5 times, and the average result was taken as represented in Table 1.

The molybdenum blue method [26] is used to measure the adsorption efficiencies of the solutions via UV–visible spectrophotometry. The following Eqs. 1 and 2 are used for the calculations.

$$P_t = W\left(C_i - C_f\right) \times m \quad (1)$$

$$\text{Ade} = \frac{C_i - C_f}{C_i} \times 100 \quad (2)$$

where

P_t (mg/g) is the amount of phosphate absorbed by the beads;
C_i and C_f (mg/L) represent the initial concentration and final concentrations of phosphate;
W denotes phosphate solution in volume and m (g) is the weight of the adsorbent mass.

3 Results and Discussion

3.1 *FE-SEM*

Figure 3 shows the FE-SEM pictures of the synthesized composite beads of La-Bnt@polysulfone. As shown in the figure, the polysulfone composite bead is smooth

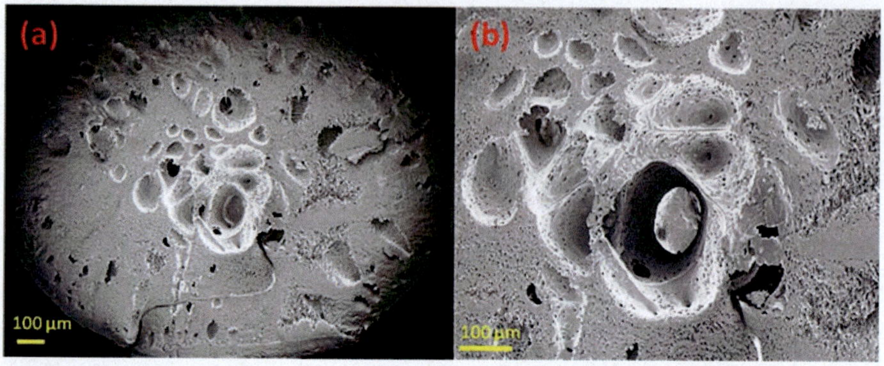

Fig. 3 **a** FE-SEM images of the cross section of the bead before adsorption and **b** high magnification view of La-Bnt@polysulfone bead showing the pore channels

and porous. It can adsorb dye and phosphate ions. Figure 3a shows the cross-sectional image of the bead, and Fig. 3b is a higher magnified image of it showing the uncountable number of channels for feed material to pass through and impurity material (phosphate and dyes) to get adsorbed inside. The bigger size pore channels will allow a high amount of feed to enter and less amount of impurity to adsorb and the small channels will let a small amount of feed enter but high adsorption of impurity will occur. The cross-sectional images demonstrate that the number of small channel pores is very high in numbers.

3.2 *FTIR*

FTIR spectroscopy is a powerful tool with a broad variety of applications, but data interpretation isn't always straightforward. Nature's absorbed energy response determines the complete spectrum that is created (hence the Fourier transform portion of the name). In the spectrum, the absorbed bands are deteriorating and becoming just vaguely distinguishable. For the FTIR data using the same dope solution, the film has been made, and after that, the analysis was done (because we tried bead for FTIR analysis but we failed so after making the film the analysis has been done and reported here). In Fig. 4, the C–H bending vibrational absorption is responsible for the peak at 1260 cm^{-1}. The characteristic peaks of polysulfone for symmetric stretching of O = S = O, asymmetric stretching of O–C–O, and the aromatic ring stretchings at 1484, 1410, 1619, and 1567 cm^{-1} are visible in the spectrum. The FTIR spectra of polysulfone absorbed heavy materials ions in La–Bnt show that some of the absorption peaks following the absorption of heavy metal ions are offset. The Fermi resonance effect is degraded by certain comparable absorption peaks. The Si–O–Si group band (~ 803 cm^{-1}) root in bentonite confirmed its presence in the polymer beads [27].

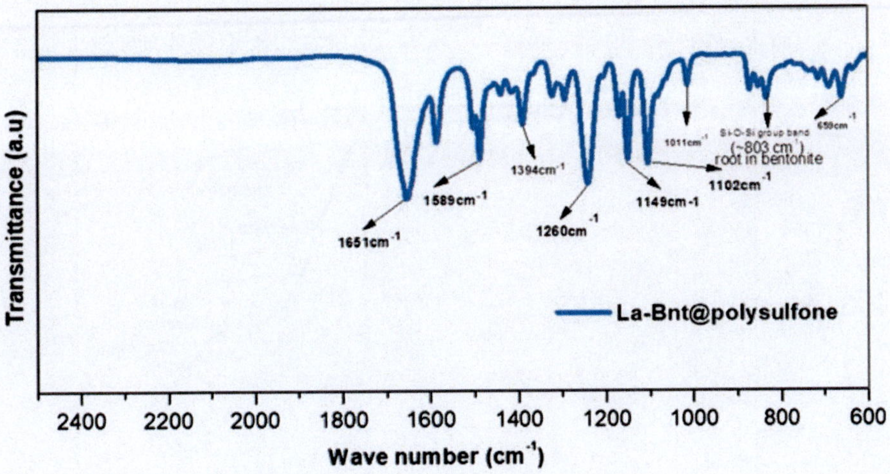

Fig. 4 ATR-FTIR spectra of La(III)-bentonite@polysulfone composite

3.3 *TGA*

La(III)-bentonite@polysulfone beads were characterized by using TGA analysis as shown in Fig. 5. The thermal stability of these beads was checked. La(III)-bentonite@polysulfone beads show a considerable variation in thermal properties. In Fig. 5, three major weight losses were found in the beads. There was gradual weight loss observed in La(III)-bentonite@polysulfone beads, i.e., the first weight loss was observed in the temperature range of 0–150 °C due to some volatile substances and the evaporation of water molecules. For the polymeric decomposition in La(III)-bentonite@polysulfone beads, second weight loss was observed at $\geq$ 180 °C. In the membrane, the third weight loss was found at 550–650 °C and above due to the decomposition of metallic and aromatic compounds of La(III)-bentonite@polysulfone beads. The decomposition process in La(III)-bentonite@polysulfone beads is moderately reduced by the addition of nanoparticles due to ligand undergoing decomposition from La-Bnt which lies at 300–530 °C.

3.4 *Performance of La(III)-Bentonite@Chitosan and La(III)-Bentonite@Polysulfone Composite Beads*

For the investigation of dye and phosphate removal efficiencies, the preliminary experiments were performed with La(III)-bentonite@chitosan and La(III)-bentonite@polysulfone composite beads and the results are shown in Fig. 6. After the loading of La(III)-bentonite in chitosan and polysulfone composite beads the dye and phosphate removal efficiency was dramatically increased, indicating that

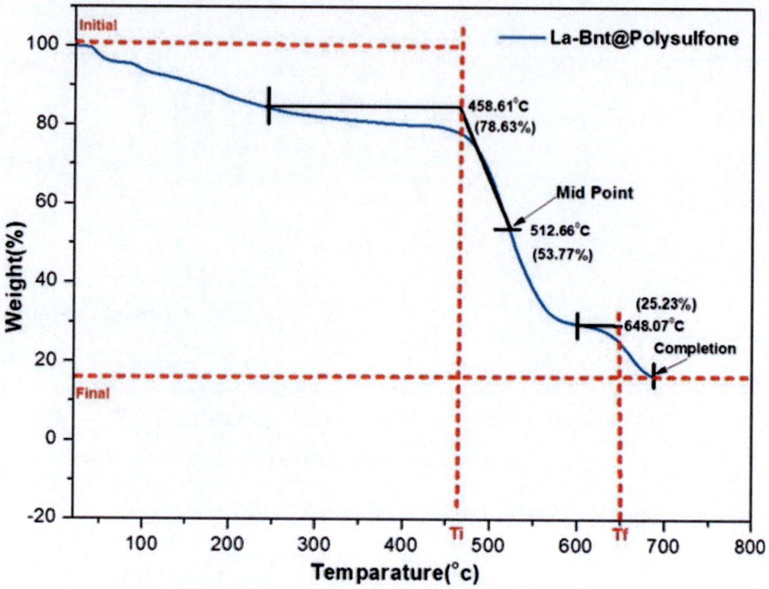

Fig. 5 TGA of La(III)-bentonite@polysulfone composite beads

La(III)-bentonite play an important role in enhanced phosphate and dye affinity. As a result, 19.82 and 19.54 mg/g adsorption capacity were obtained by La(III)-bentonite@chitosan and La(III)-bentonite@polysulfone beads respectively with phosphate concentration = 20 mg/L. The added La(III)-bentonite nanoparticles provide more positive charges on the beads and promote the removal of phosphate. Moreover, the La(III)-bentonite@chitosan beads showed the highest phosphate removal efficiency. The MV dye with a concentration of 30 mg/L was tested with the same beads and found the adsorption in the range of 79.2–99.4%. Based on these results, all further studies are going on in detail. The comparative study with other research results has been shown in Table 2. It is also observed that the obtained results are good and continuous work is going on, on this topic in our group. We hope after doing a detailed study we will be able to remove low to high concentrations of phosphate and dye molecules.

4 Conclusion

La(III)-bentonite@chitosan and La(III)-bentonite@polysulfone composite beads were successfully prepared by the phase inversion/separation method. The prepared beads have shown the best results for the removal of methyl violet and phosphate. At

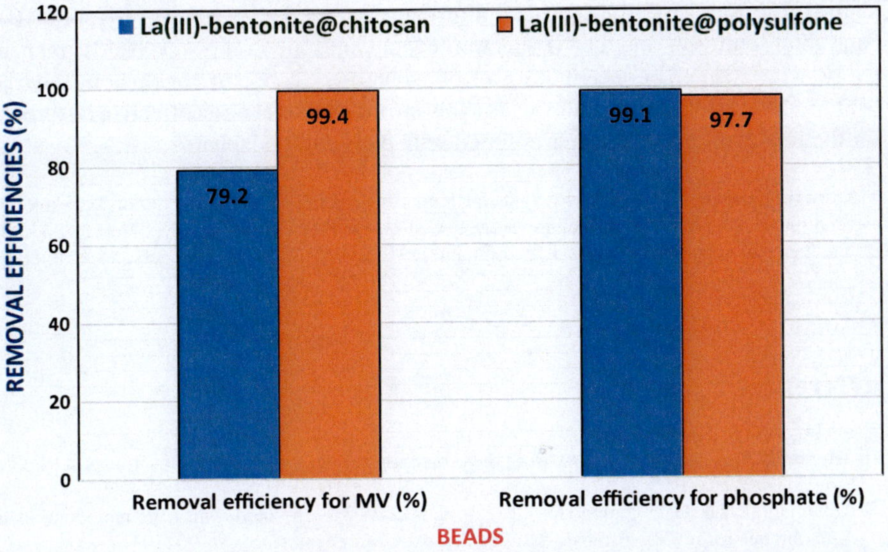

Fig. 6 Phosphate removal efficiencies of the La(III)-bentonite@chitosan and La(III)-bentonite@polysulfone composite beads (MV dye concentration 30 mg/L and, initial phosphate concentration = 20 mg/L, nanoparticles loading = 0.1 g, pH = 7)

Table 2 Phosphate adsorption capacities of La(III)-bentonite@chitosan and La(III)-bentonite@polysulfone composite beads and other reported adsorbents

Adsorbent	Initial phosphate concentration (mg/L)	Adsorption capacity (mg/g)	References
Chitosan/Ca-OMMT beads	100	75.4	[13]
Mixed La/Al pillared montmorillonite (LaAl-PILC)	5	13	[21]
A tetra-amine Cu(II) (TAC@CS) chitosan beads	100	41.4	[28]
Chitosan hydrogel beads	100	28.9	[29]
Montmorillonite (diameter of the particles is < 0.1 mm)	5	0.13	[30]
Cross-linked chitosan bead	100	52.1	[31]
Montmorillonite	3.5–25	0.7	[32]
La(III)-bentonite@polysulfone beads	20	19.82	This work
La(III)-bentonite@chitosan beads	20	19.54	This work

neutral pH (pH = 7), the bead's adsorption capacity was 99.1% and 97.7% for La(III)-bentonite@chitosan and La(III)-bentonite@polysulfone composite beads, respectively, with 20 mg/L concentration of phosphate at 25 °C. In conclusion, La(III)-bentonite@chitosan composite beads show strong potential as an effective adsorbent for the treatment of wastewater polluted with dyes and phosphates.

Acknowledgements The authors give special thanks to Director, CSIR-NEIST for his keen interest and continuous support. Authors acknowledge to CSIR-New Delhi, India, and DST New Delhi India for the financial support under the OLP-2064 and DST/NM/NT/2018/143 projects, respectively.

References

1. Boretti A, Rosa L (2019) Reassessing the projections of the World Water Development Report. npj Clean Water 2:15
2. Zhao Q, Zhao DL, Chung TS (2020) Nanoclays-incorporated thin-film nanocomposite membranes for reverse osmosis desalination. Adv Mater Interfaces 7:1902108
3. Amy G, Ghaffour N, Li Z, Francis L, Linares RV, Missimer T, Lattemann S (2017) Membrane based seawater desalination: present and future prospects. Desalination 401:16–21
4. Mateo-Sagasta J, Zadeh SM, Turral H, Burke J (2017) Water pollution from agriculture: a global review. Executive summary. FAO, Rome, Italy; International Water Management Institute (IWMI), CGIAR Research Program on Water, Land and Ecosystems (WLE), Colombo, Sri Lanka, p 35
5. Karki S, Ingole PG (2022) Development of polymer-based new high performance thin-film nanocomposite nanofiltration membranes by vapor phase interfacial polymerization for the removal of heavy metal ions. Chem Eng J 446:137303
6. Fallahnejad Z, Bakeri GH, Ismail AF (2022) Functionalized halloysite nanotubes incorporated thin film nanocomposite nanofiltration membrane for treatment of wastewaters containing metal ions. Process Saf Environ Prot 157:334–351
7. Yadav KK, Singh DK, Anitha M, Varshaney L, Singh H (2013) Studies on separation of rare earths from aqueous media by polyethersulfone beads containing D2EHPA as extractant. Sep Purif Technol 118:350–358
8. Gong XC, Luo GS, Yang WW, Wu FY (2006) Separation of organic acids by newly developed polysulfone microcapsules containing trioctylamine. Sep Purif Technol 48:235–243
9. Zhao C, Xue J, Ran F, Sun S (2013) Modification of polyethersulfone membranes—a review of method. Prog Mater Sci 58:76–150
10. Karki S, Gohain MB, Yadav D, Thakare NR, Pawar RR, Hazarika S, Ingole PG (2023) Building rapid water transport channels within thin-film nanocomposite membranes based on 2D mesoporous nanosheets. Desalination 547:116222
11. Van den Berg C, Roelands CPM, Bussmann P, Goetheer ELV, Verdoes D, Wielen LAMVD (2009) Preparation and analysis of high capacity polysulfone capsules. React Funct Polym 69:766–770
12. da Silva Alves DC, Healy B, Pinto LADA, Cadaval TRS, Breslin CB (2021) Recent developments in chitosan-based adsorbents for the removal of pollutants from aqueous environments. Molecules 26:594
13. Jang J, Lee DS (2019) Effective phosphorus removal using chitosan/Ca-organically modified montmorillonite beads in batch and fixed-bed column studies. J Hazard Mater 375:9–18
14. Roy M, Saha R (2021) Dyes and their removal technologies from wastewater: a critical review. Intell Environ Data Monit Pollut Manage 127–160

15. Yan J, Yuan W, Liu J, Ye W, Lin J, Xie J, Huang X, Gao S, Xie J, Liu S, Chen W, Zhang H (2019) An integrated process of chemical precipitation and sulfate reduction for treatment of flue gas desulphurization wastewater from coal-fired power plant. J Clean Prod 228:63–72
16. Liu HL, Chiou YR (2006) Optimal decolorization efficiency of reactive red 239 by UV/ZnO photocatalytic process. J Chin Inst Chem 37:289–298
17. Alfonso-Muniozguren P, Lee J, Bussemaker M, Chadeesingh R, Jones C, Oakley D, Saroj D (2018) A combined activated sludge-filtration-ozonation process for abattoir wastewater treatment. J Water Process Eng 25:157–163
18. Ingole PG, Sawant SY, Ingole NP, Pawar RR, Bajaj HC, Singh K, Cho MH, Lee HK (2016) Preparation of activated carbon incorporated polysulfone membranes for dye separation. Membr Water Treat 7:477–493
19. Gohain MB, Yadav D, Karki S, Yadav A, Thakare NR, Hazarika S, Lee HK, Ingole PG (2022) Development of Antifouling Thin-Film Composite/Nanocomposite Membranes for Removal of Phosphate and Malachite Green Dye. Membranes 12:768.
20. Li W, Mu B, Yang Y (2019) Feasibility of industrial-scale treatment of dye wastewater via bio-adsorption technology. Biores Technol 277:157–170
21. Tian S, Jiang P, Ning P, Su Y (2009) Enhanced adsorption removal of phosphate from water by mixed lanthanum/aluminum pillared montmorillonite. Chem Eng J 151:141–148
22. Kong L, Tian Y, Li N, Liu Y, Zhang J, Zhang J, Zuo W (2018) Highly-effective phosphate removal from aqueous solutions by calcined nano-porous palygorskite matrix with embedded lanthanum hydroxide. Appl Clay Sci 162:507–517
23. Kuroki V, Bosco G, Fadini P, Mozeto A, Cestari A, Carvalho W (2014) Use of a La (III)-modified bentonite for effective phosphate removal from aqueous media. J Hazard Mater 274:124–131
24. Zhang Y, Xu Y, Cui H, Liu B, Gao X, Wang D, Liang P (2014) La(III)-loaded bentonite/chitosan beads for defluoridation from aqueous solution. J Rare Earth 32:458–466
25. Younes I, Rinaudo M (2015) Chitin and chitosan preparation from marine sources. Struct Prop Appl Mar Drugs 13:1133–1174
26. Rotzetter A, Kellenberger C, Schumacher C, Mora C, Grass R, Loepfe M, Luechinger N, Stark W (2013) Combining phosphate and bacteria removal on chemically active filter membranes allows prolonged storage of drinking water. Adv Mater 25:6057–6063
27. Xu X, Cheng Y, Wu X, Fan P, Song R (2020) La(III)-bentonite/chitosan composite: a new type adsorbent for rapid removal of phosphate from water bodies. Appl Clay Sci 190:105547
28. Kumar IA, Viswanathan N (2018) Preparation and testing of a tetra-amine copper(II) chitosan beads system for enhanced phosphate remediation. Carbohydr Polym 183:173–182
29. Dai J, Yang H, Yan H, Shangguan Y, Zheng Q, Cheng R (2011) Phosphate adsorption from aqueous solution by disused adsorbents: chitosan hydrogel beads after the removal of copper(II). Chem Eng J 166:970–977
30. Tilková A, Majorová P, Seidlerová J (2013) Using montmorillonite for phosphate removal. Adv Sci Focus 1:351–353
31. Mahaninia MH, Wilson LD (2017) Phosphate uptake studies of cross-linked chitosan bead materials. J Colloids Interfaces Sci 485:201–212
32. Edzwald JK, Toensing DC, Leung MC-Y (1976) Phosphate adsorption reactions with clay minerals. Environ Sci Technol 10:485–490

Ultra-Precise Single-Point Diamond Turning Process and Its Low-Cost Alternative Methods

Borad M. Barkachary and Shrikrishna Nandkishor Joshi

1 Introduction

One of the ultra-precision material removal processes is single-point diamond turning (SPDT). It uses single crystal or poly-crystal diamond as cutting tools with an incredibly sharp edge to create components with A° level surface finish [1]. The main distinction between a standard lathe and an SPDT machine is that the latter employs a fine diamond-tipped bit (cutting-edge radius between 20 and 100 nm) for its machining, hydrostatic/aerostatic bearings for the table and spindle, granite bed for vibration isolation and feedback system with sub-nanometer resolution. An illustration of the SPDT machine is shown in Fig. 1.

The studies on single crystal diamond turning were first documented by Jesse Ramsden in 1779 [2] in which a screw for separating engines made of hardened steel was cut with a diamond. Using this as a foundation, the ultra-precision machining (UPM) was initially developed in the 1960s to create super-polished surfaces with sub-micrometric form accuracy and surface waviness [3]. Then, in the late 1970s, the SPDT process was made available for commercial use. Since then, it has evolved into a key process in the precise manufacture of infrared imaging systems, spherical, aspherical and freeform optics, precision moulds, metal mirrors, reflectors and lenses needed for cameras, binoculars, projectors and human vision.

B. M. Barkachary (✉)
Department of Mechanical Engineering, Jorhat Institute of Science and Technology, Sotai, Jorhat, Assam 785010, India
e-mail: borad.barkachary@gmail.com

S. N. Joshi
Department of Mechanical Engineering, Indian Institute Technology Guwahati, Guwahati, Assam 781039, India
e-mail: snj@iitg.ac.in

S. N. Joshi et al. (eds.), *Low Cost Manufacturing Technologies*,
https://doi.org/10.1007/978-981-19-8452-5_8

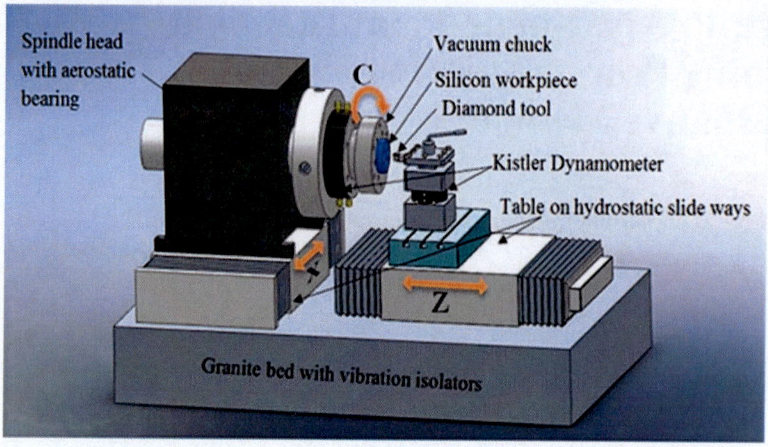

Fig. 1 Schematic CAD diagram of SPDT machine

1.1 Applications of SPDT

A few of the components produced with SPDT are shown in Fig. 2. Optics, instrumentation, electronics, military, medical and space industries all benefited from the use of SPDT. Here is a list of them [4–7]:

- Reflective mirrors,
- Precision mould inserts and cores for camera, binoculars and projectors lenses.
- Telescopic mirrors for astronomy.
- Lenslet arrays, spherical and aspherical optical lenses (Diffractive, Fresnel, Toric and Free form optics lenses).
- Medical instrumentation, contact and scleral lenses for human vision and bio-implants.
- Laser optics and infrared hybrid lenses for thermal imaging, missile guidance systems for defense sector.

1.2 Challenges of Conventional Brittle Materials Machining

Brittle materials are difficult to cut mechanically due to the occurrence of fracture without appreciable plastic deformation. Compared to ductile materials, brittle materials possess extremely small fracture strain than their yield strength. As such, the tensile and compressive stresses exerted by the tool originate small cracks in the workpiece. Through the path of least resistance, the initial fracture swiftly spreads to create discontinuous chips. The brittle material fails to withstand the load before it undergoes yielding, and the chips are formed by the onset and propagation of cracks. Thus, brittle material generates discontinuous chips, which have a major impact

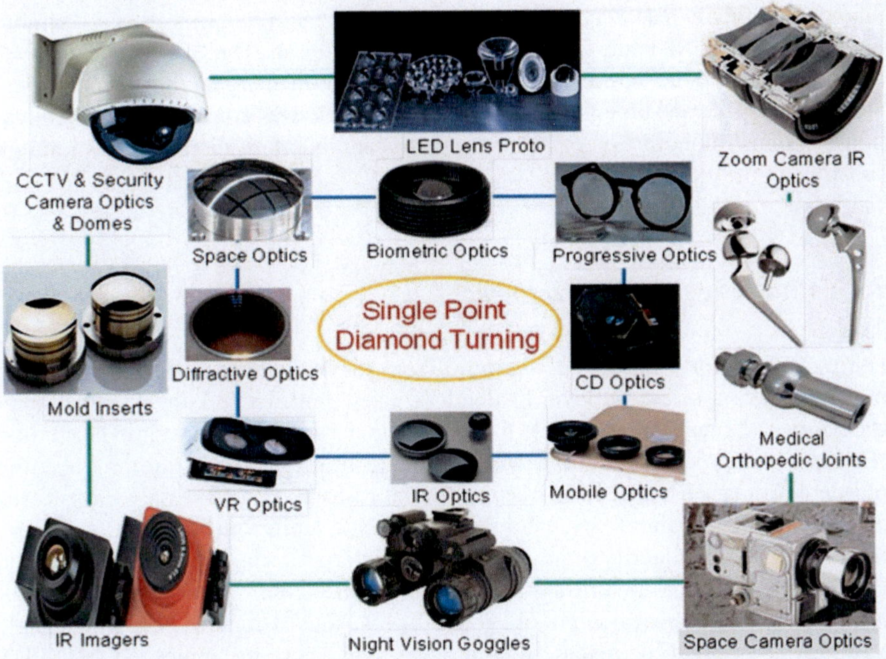

Fig. 2 SPDT-based components and products (Recreated from Optics and Allied Engg. Pvt. Ltd., India, [8])

on the surface integrity of the machined parts. To supersmooth the surfaces, traditional grinding and polishing processes are used; however, these are time-consuming, laborious, reliant on skilled labour and unproductive [9].

Mechanical processing of Brittle material generates fragmented chips owing to brittle failure at the shear plane prior to actual plastic flow. Brittle fracture during machining has a substantial impact on the surface integrity and dimensional accuracy of the machined part. As a result, conventional machining is not often suggested for producing precision components with elevated surfaces on brittle materials. As per the literature, ductile mode machining of brittle materials can be achieved if the depth of cut is maintained less than the critical depth (nm). A variety of challenges arises during micro-nano-machining due to the nanometric depth. Tool tip loses their sharpness and become rounded when the depth of cut decreases to nanometric depth, i.e. less than the cutting-edge radius, resulting in a very negative effective rake angle tool. Materials can no longer be considered homogeneous and isotropic, and the cutting process is dominated mainly by ploughing, extrusion, rubbing or sliding action rather than sharing.

The roots of diamond turning may be traced back to the 1930s, when the watch dial components were first manufactured using diamond turning by the jewellery industry [10]. However, the use of mono-crystal diamond tools for SPDT of aluminium and beryllium copper dates back to 1966 [6]. The researchers then began working on diamond turning in research laboratories in the USA, including Lawrence Livermore National Lab (LLNL) and Oak Ridge Y-12 National Lab. The SPDT machine was commercialized in the late 1970s and since been used in energy, optical, computing, electronics, space and defence sectors [3]. This process was employed in the production of drums in photocopiers, scanner mirrors and memory discs from aluminium substrate in the 1980s.

1.3 Challenges in SPDT Process

During the past forty years, experts from all over the world have identified that brittle materials have a threshold layer or depth of a few nm in size known as the "ductile regime", beneath which it acts like ductile materials. Ductile regime machining (DRM) is the process of cutting where chips of brittle materials are produced by plastic deformation rather than fracture. Initial research on DRM on rock salt was published by King and Tabor, who suggested that significant hydrostatic pressure is essential for the ductile removal of brittle material during frictional wear [11]. Plastically deforming the brittle material can produce continuous ductile chips if it is machined below critical depth of cut [12] or by using extreme negative rake angle tool [13] or external hydrostatic pressure [14, 15]. These principles aid the SPDT process in suppressing the early crack initiation required for nanometric ductile mode machining of brittle materials. Moreover, appropriate selection of regulated parameters, viz. cutting speed, feed rate, rake angle, depth of cut, and kind of coolant is also crucial [12].

1.4 Process Parameters of SPDT

The SPDT process's product quality and process performance are affected by several aspects. Process parameters, cutting tool shape, workpiece and tool material qualities (microstructure and crystal orientation), machine tool settings and machining conditions are among them. These parameters are widely classified as machining process parameters, tool geometry parameters, workpiece-tool material characteristics and other machining condition parameters. It is important to understand the implications of these factors on SPDT performance measures such as phase change, machining (cutting and thrust) forces, chip structure, surface roughness and tool wear. Figure 3 displays the SPDT process influencing elements that must be investigated in order to boost product quality and process efficiency.

Significant scientific effort on experimental investigations of the SPDT process on diverse materials is reported in the literature. However, the scope of SPDT research is constrained by its high equipment and operational expenditure. Numerical methods are the next option to be adopted to carry out rigorous study of the process. An exhaustive literature analysis on all elements of the SPDT process revealed that there

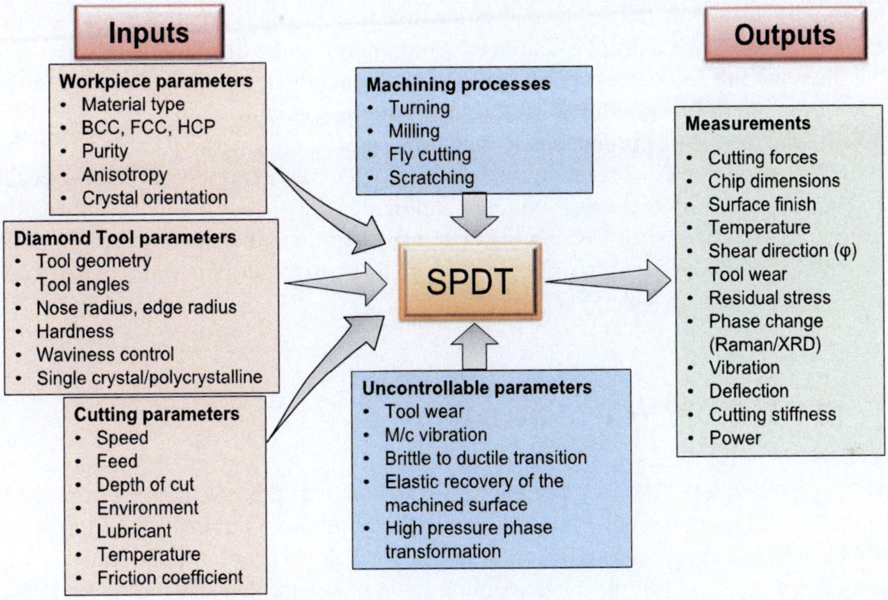

Fig. 3 Inputs and output parameters during SPDT process

is relatively little information on numerical modelling of the SPDT process. Furthermore, relatively little research on the parametric optimization of SPDT process to optimise the process parameters to increase process efficiency and product quality has been documented.

1.5 Motivation for the Present Work

Brittle materials are extremely difficult to process using traditional material removal methods due to their brittleness and poor fracture toughness. As a result, it is susceptible to fracture and subsurface damage during the machining process. Machining soft and ductile materials is also problematic due to the creation of built up edges (BUE) and accompanied with burnishing, elastic recovery, plastic deformation and materials swelling [16]. The chemical affinity of diamond with carbon produces significant tool wear when machining ferrous materials. The formation of highly localised stresses, strains and temperature adds to the complexity. For this reason, it is crucial to comprehend how the SPDT's process, material and tool characteristics affect its performance parameters, such as machining forces and surface roughness, in order to enhance product quality and process effectiveness. Although optical-finish surfaces can be generated using SPDT, it is crucial to carry out physical experiments to fully grasp the effects of various process parameters, tool geometry, workpiece material, working condition on process output and product quality. However, because to the nanometric level material removal, they are time intensive and expensive. This served as inspiration for the current research in comprehending the machining of brittle and ductile materials by means of low-cost alternative method. Numerical simulation is one of the alternative methods to study the nanometric cutting processes and their influencing factors on product quality and process efficiency without carrying out the actual experiments. The present work focuses upon the short review of SPDT process and its numerical simulations for understanding of SPDT process and its influencing factors on process outputs. Researchers and industrial engineers are expected to be benefited from the information gained through this current work in order to carry out effective and high-quality SPDT operations.

2 SPDT of Different Materials

This paper presents a detailed review on SPDT and its alternative method of process modelling and simulation. Figure 4 shows the number of publication reported on SPDT of various engineering materials (till 2018). Global scholars have made significant attempts to comprehend machining mechanics, particularly DRM of brittle materials to achieve crack-free optical-finish surfaces by processing it in ductile mode. Researchers have performed experimental as well as numerical works on SPDT of various materials. In that, Silicon is the material that is studied most due to its growing demand in industries.

Literature also reports machining of non-ferrous ductile metals such as aluminium, copper PMMA, nickel, brass for key components of telescopes, video projectors, lasers and imaging systems [17]. Brittle materials such as silicon carbide, germanium and glass have also been machined in the ductile mode by accurately managing the

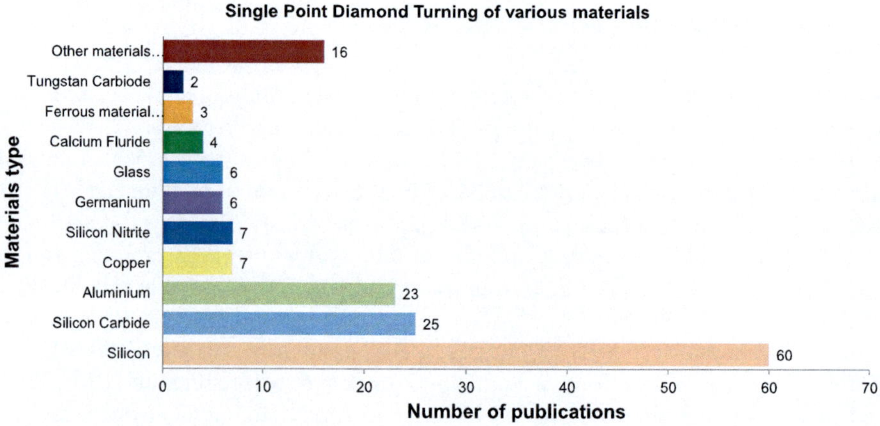

Fig. 4 Number of publications reported on SPDT of various engineering materials (up to 2018)

Table 1 Published literature on SPDT of various materials

S. No.	Material	References
1	Aluminium	[18–20]
2	Copper	[21–23]
3	PMMA	[24]
4	Nickel	[25]
5	Brass	[26, 27]
6	Silicon	[12, 13, 28–32]
7	Silicon carbide	[33–38]
8	Germanium	[39–43]
9	Glass	[44–46]

process parameters, viz. depth of cut, speed and feed precisely. Some of these studies reported in the literature are enlisted in Table 1.

Based on the selected papers taken for review, three materials (viz. Silicon, Silicon carbide and Al6061) have been chosen and a comparative bar chart has been prepared to get the idea of the current worldwide scenario of SPDT process. This is shown in Fig. 5, in that, USA is the leading country for carrying out majority of research work on SPDT studies on the selected materials.

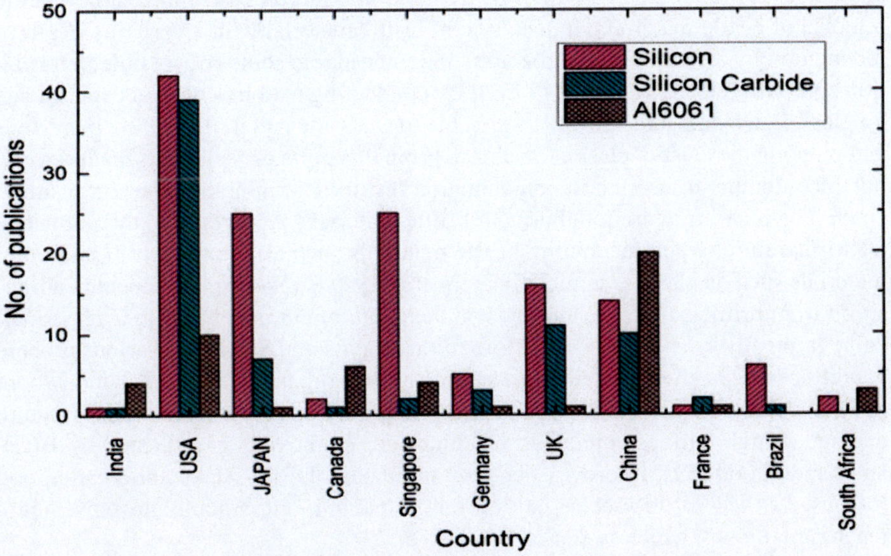

Fig. 5 Number of publications reported country wise on single-point diamond turning up to 2018

2.1 Ductile Regime Machining

Brittle materials have very low-density defects, which serve as the starting points for fracture. So, if the tool's stress field is low, it reduces and, in certain cases, terminates the nucleation of the fracture. That is, if the depth of cut is maintained below the critical depth, the stress field created at the cutting zone is likewise small, avoiding the onset of fracture at the defects. The critical thickness is the point at which all brittle materials change their behaviour from brittle to ductile. This critical thickness is governed by the depth of cut, tool rake angle, cutting-edge radius and hydrostatic pressure. If the depth of cut exceeds critical thickness, stress field increases causing the cracks to propagate from defects to form brittle chips. Each material has a different critical layer or thickness at which this transition takes place. The critical depth of cut (d_c) can be calculated by Eq. (1) [47, 48]

$$d_c = 0.15(E/H)(K_c/H)^2 \tag{1}$$

where E is the Young modulus (GPa), H is the Vickers hardness (GPa) and K_c is the fracture toughness (MPa $\text{m}^{1/2}$) of the material.

To manufacture surfaces with optical quality on brittle materials, right machining conditions with nanometric order of machining paramerter must be used. In their groundbreaking study on DRM using plastic deformation on rock salt, King and Tabor [11] hypothesised that high hydrostatic pressure is essential for the ductile removal of brittle materials. Indentation of brittle material with a very tiny depth of indentation by a sharp pointed diamond indenter leaves some irreversible deformation zone without any fracture [49–52]. By conducting nanoindentation experiments on glass, Lawn and Wilshaw [49] were able to study the material's ductile behaviour and pinpoint the elastic–plastic transition. From that point on, scientists made several notable attempts to use the nanoindentation test to determine the phase transition, crack length and fracture toughness of brittle materials. According to the published DRM literature, when machining brittle materials such as ceramics [53] and other materials such as glasses, semiconductors and crystals [54–56], there occurs a transition from brittle-to-ductile mode when the depth of cut is reduced to a very small value (typically < 1 μm). Plastic deformation of glass and SiC under various process conditions has been documented in studies by Huerta and Malkin [57] and Moore and King [58]. Brittle material machining is possible without subsurface fracture damage if high stiffness precision machines are employed, as indicated by Blake and Scattergood [59]. In order to better comprehend DRM, Blake and Scattergood [60] used SPDT to do precise cutting on germanium and silicon and presented a machining model, which is depicted in Fig. 6.

It illustrates the tool's projection perpendicular to the cutting direction, where R is the tool nose radius, d is the depth of cut, f is the feed rate, y_c is the average surface damage depth, and Z_{eff} is the distance between the tool centre and the crack propagation on the uncut shoulder. The authors claimed that ductile regime machining

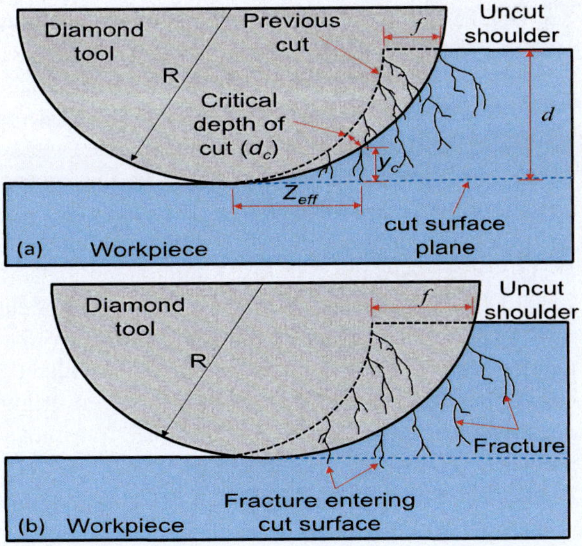

Fig. 6 Schematic of ductile regime machining (Redrawn from [59])

of brittle material is accomplished by careful selection of feed rate, depth of cut and tool nose radius, such that the average damage depth does not enter the finished cut surface.

The literature displays notable research studies on the phenomena of brittle–ductile transition during indentation [61–64], scratching [65], taper turning [65–69], scribing [70], grinding [71–73] and machining [31, 33, 37, 55, 74–77]. During a Hertzian indentation test, Lawn et al. [78] discovered ductile behaviour of brittle materials. Based on the density of defects in the brittle materials, Nakasuji et al. [79] described the ductile transition and plastic deformation. Similar findings have been made by Yan et al. [14] and Bridgeman and Šimon [80] showing that at ambient temperature strong hydrostatic pressure may promote plastic flow in silicon and glass. As a result, brittle materials including diamond can be deformed plastically even at room temperature when subjected to high hydrostatic stresses. According to Morris et al. [81], phase transition brought on by a high degree of hydrostatic pressure/stress is what causes the ductile behaviour, or plastic flow of brittle material. A generalised theory for the brittle-to-ductile transition in Si and $LiNbO_3$ micro-machining and micro-indentation was put out by Shimada et al. [82]. It has been proposed that any material may be machined in ductile mode, regardless of ductility, if it is machined at sufficiently small-scale depth, i.e. nanometric level. According to Leung et al. [12] to achieve a good-quality surface using SPDT, the chip thickness should be on the range of a few micrometres (below its critical value); i.e. the machining process should be in the ductile regime. Fang and Venkatesh [56] achieved tens of nanometer surface finish on single crystal silicon using diamond cutting, and a ductile–brittle transition thickness of 236 nm was obtained. The authors found that zero-degree rake angle tools produce a higher grade surface finish than negative rake angle tools. In contrast, Patten [40] found that negative rake tools outperform zero and positive degree rake angle tools in the ductile mode machining. For similar reason, Yan et al. [14] and Yoshino et al. [15] performed machining tests on silicon under a high external hydrostatic pressure (more than 400 MPa) and noticed yielding of continuous chips with smooth work surfaces. Patten and Gao [13] observed that cutting with both the rake (–45°) and the clearance face (–85°) yielded a flawless ductile cut with no sign of surface damage. Furthermore, they noted that the hydrostatic pressure created primarily by the negative rake angle tool is critical in minimising fracture and producing smooth surfaces. Using large tool feeds on the order of 20 m/rev, Yan et al. [83] assessed the viability of ductile regime silicon machining. The impact of varying the tool-edge radius on the effectiveness of cutting single crystal silicon was investigated by Fang and Zhang [69]. The study demonstrated that ductile regime cutting could be carried out even with a tool with a 0° rake angle if the undeformed chip thickness was below a certain value, converting the effective rake angle to a negative rake angle. The ductile to brittle transitions (DBTs) for 6H-SiC, 4H-SiC and CVD 3C-SiC, respectively, are 70, 820 and 550 nm, according to research by Bhattacharya et al. [84], Patten et al. [85] and Ravindra and Patten [86] that looked at the critical depth of cut at which the DBT occurs. An in-depth analysis of DRM on semiconductors, ceramics and glass was presented by Kovalchenko [87]. Hatefi and Abou-El-Hossein [88] also provided a comprehensive review on SPDT in terms of producing optical surface finish.

3 Numerical Modelling and Simulation of SPDT Process

Gradually, it is possible to see the use of numerical methods like molecular dynamics (MD) and finite element method (FEM) as a worthy substitute for the costly, laborious and time-consuming physical experiments. As the computer systems have advanced rapidly, numerical analysis has become a potent tool to examine machining processes and providing a scientific understanding of how materials respond to machining. From conventional to micro- to nanoscale machining, numerical models can anticipate the behaviour of the material that is sometimes difficult to measure physically.

Important research on numerical modelling of SPDT to replicate the complicated physical phenomena is reported in the literature. Till now, many numerical simulation methods have been reported by many researchers such as finite element method (FEM), finite difference method (FDM), finite volume method (FVM), smoothed particle hydrodynamics (SPH), extended FEM (XFEM), molecular dynamics (MD) and their hybrids. All of the methods come with their advantages and disadvantages. These methods can shed light on the effects of cutting process that are sometimes difficult to see through experimentation. According to reports, researchers used finite element method (FEM) and molecular dynamics (MD) techniques in particular to describe and simulate the UPM process. FEM employs the concept of continuum

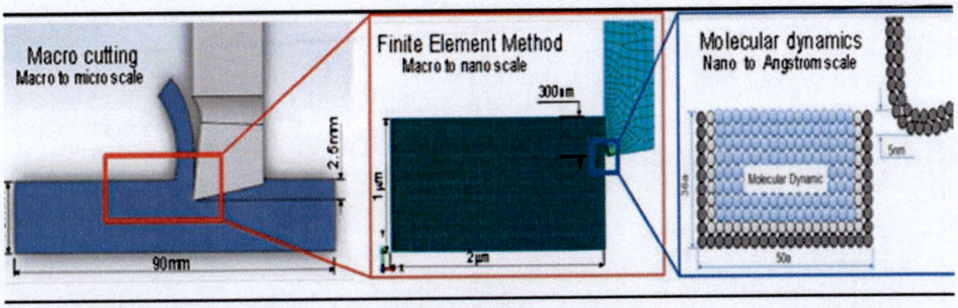

Fig. 7 Dimensional comparisons of experimental macro cutting, FEM and MD simulations

mechanics, whereas MD employs discrete mechanics, which takes into account inter-atomic bonding of atoms. Dimensional comparisons of macro-, FEM and MD simulations are shown in Fig. 7. While the highest dimension that can be taken into account in an MD simulation is a few nanometers, the working dimension in FEM may range from macro- to nanoscale.

3.1 Molecular Dynamics Simulation of SPDT

The most acknowledged advanced numerical simulator that can replicate the circumstances of nanometric cutting is MD simulation. It has the potential to simulate at the atomic level and to characterise the evolution of the material's microstructure that led its adoption by several scholars. Alder and Wainwright presented the pioneering work on molecular dynamics (MD) simulation in the late 1950s [89]. Later, in the 1980s, Lawrence Livermore National Laboratories (LLNL), USA, adopted the MD to imitate UPM [90]. Several researches on various facets of nanometric machining have been conducted since then. These include crystallographic orientation, tool-edge radius and minimum depth of cut on the chip formation mechanism, phase transition, defects in the workpiece material and diamond tool wear. Using MD, Liu et al. [91–93] examined the scratching process of SiC. It was reported that the phase shift of SiC to its amorphous phase triggered ductile regime scratching of SiC at the nanoscale depth of cut. Wang et al. [94] used the model based MD approach to simulate the scratching of SiC hard particles with a diamond tool in order to study the groove wear on the tool's flank face.

Figure 8a shows MD simulations of the nanometric chip removal technique used in copper micro-cutting by Shimada and Ikawa [95]. This study revealed that the expected chip morphologies, cutting forces and specific energy were in good accord with the experiments. Renormalized molecular dynamics (RMD) simulations of silicon machining were reported by Inamura et al. [96] in order to study the crack formation mechanism. Though silicon shows a brittle-ductile transition based on the scale of machining under normal atmospheric conditions, the findings indicate that it may be machined in a ductile mode in an absolute vacuum environment.

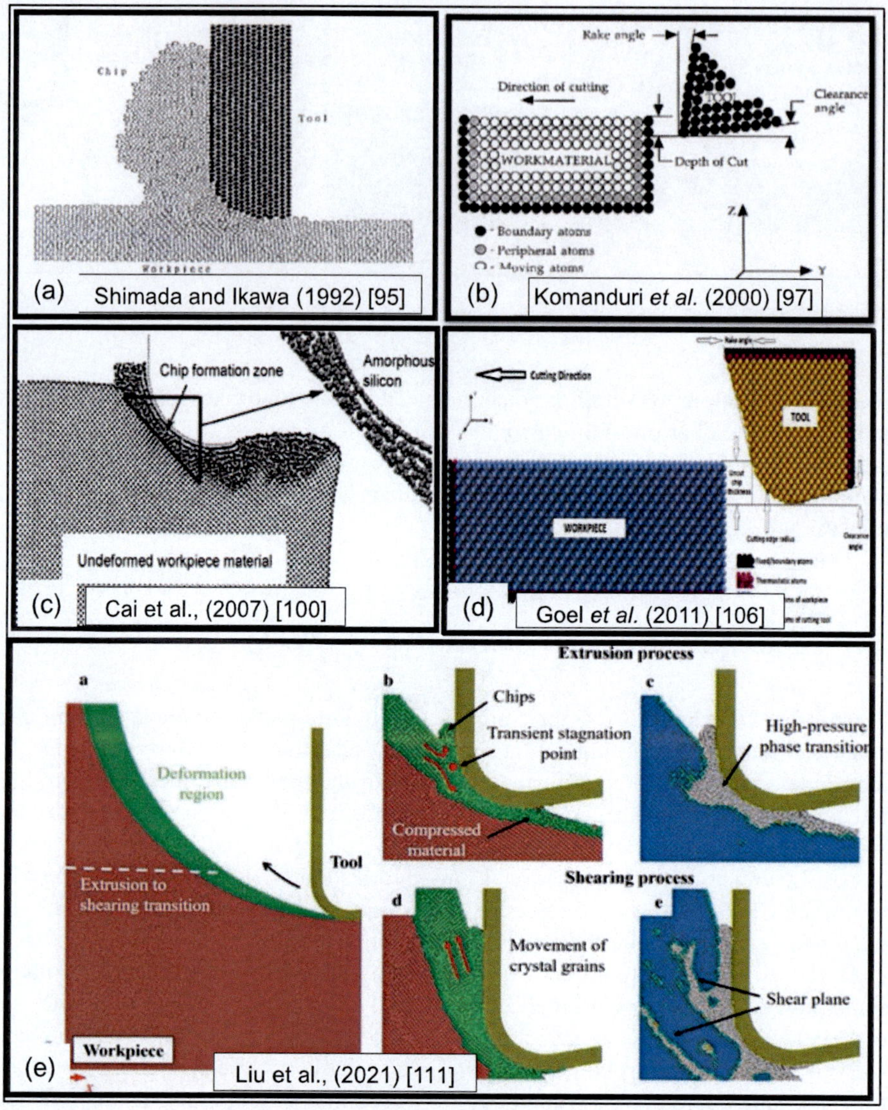

Fig. 8 MD simulation of nanometric cutting **a** Shimada and Ikawa [95], **b** Komanduri et al. [97], **c** Cai et al. [100], **d** Goel et al. [106] and **e** Liu et al. [111]

The effects of crystal orientation, cutting direction and positive rake angle on nanometric cutting of single crystal aluminium were investigated by Komanduri et al. [97] (Fig. 8b). It was discovered that the cutting forces fluctuate cyclically with crystal orientation and cutting direction. It was also found that as the rake angle increases, so does the cutting force. Likewise, Komanduri et al. [98] investigated the nature of material removal and surface generation in UPM and grinding of silicon by using the Tersoff potential to perform MD simulation while varying the depth of cut, width of cut, rake angle and clearance angle. During the investigation, four distinct material removal methods, such as compression of the work material ahead of the tool, chip formation that resembles an extrusion-like process, side flow and subsurface distortion in the machined surface, were reported. For the study of nanometric cutting processes utilising MD simulation, Cheng et al. [99] and Cai et al. [100] have made major contributions to the field. Cai et al. [100] undertook nanometric machining of silicon wafer by employing physical tests and MD simulations to investigate the process of ductile chip generation (see Fig. 8c). It was determined that phase transition, not atomic dislocation, was responsible for the plastic deformation. Additionally, by doing MD-based simulations, Komanduri and Raff [101], Oluwajobi [102], Goel et al. [103], Guo et al. [104] and Abdulkadir et al. [105] contributed significant knowledge and set a firm basis for the research of nanometric cutting processes. Goel and his colleagues presented substantial work on MD simulations of nanometric silicon and silicon carbide machining. The authors investigated the DBT of SiC [37], atomistic characteristics of SiC ductile response [106], influence of temperature and crystal orientation on tool wear [13, 107, 108] and effect of silicon microstructure on cutting behaviour [109]. Figure 8d depicts the usual MD model used by Goel et al. [106] to simulate the SPDT process. Several researchers [29, 31, 32, 110] studied the wear of diamond tools and found that during silicon machining, flank wear predominates. The development of silicon carbide at the tool-work contact during the machining of silicon was described by Goel et al. [31, 107]. These phenomena provided details about the beginning of diamond tool wear and elucidated the tribochemistry involved in the process that leads to diamond tool wear. In the recent MD study, Liu et al. [111] reported a hybrid machining of single crystal silicon with thermal and vibration assistants to investigate material removal behaviour and subsurface damage formation (Fig. 8e). It was observed that the temperature and vibration should be set carefully to suppress the vacancies to reduce the cracks and fracture in the workpiece.

3.2 *Finite Element Simulation of SPDT*

The MD approach was shown to successfully simulate the mechanism of nanocutting of brittle materials. An issue with the method is that it can only address very small continuum size as a few million atoms or even less and the time scales are in the order of picoseconds. Massive amounts of memory and computation time are needed for the MD simulation. Since the results are generated for atomistic scale, it is challenging to compare them to experimental data [112] and use them in real SPDT process conditions. The finite element method (FEM) is another alternative to this problem. FEM simulation covers a wide range of process continuums, from macro- to nanometric size, and can provide insight into cutting process characteristics that are sometimes impossible to see via tests. It has the ability to predict temperatures, stresses, strains, shear angles, chip geometry and cutting forces. FEM-based modelling may easily include the impacts of severe deformation, strain rate effect, tool–chip interactions and their friction, local heating and temperature effect and other phenomena. Besides complex physical interactions between the tool and the workpiece, a variety of boundary and loading conditions, such as thermal, structural, electrical and magnetic, can be defined and simulated. FEM has been used to simulate macro and meso-level machining process since it has been developed as a commercialization. FEM was implemented for machining simulations in the early 1970s using user defined finite element codes. Later, many commercial software were developed with in-build FE codes and user graphical interface for performing all kinds of structural, thermal and machining simulations. Significant work on macro–micro-metal cutting simulations using finite elements is described in the literature [113–116]. The published literature on general FEM machining model reported during 1976–2002 is presented in [117]. However, there has not been a lot of work done on developing an effective SPDT process model that can be used for a systematic parametric analysis to find the optimal process parameters. Figure 9 depicts the pi-chart of SPDT studies that are adopted by most of the researchers. It shows that more than 50% of the reported literature carried out experimental studies on SPDT, whereas 20% have used MD simulation to analyse the process. Only 3% and 20% of reported literature used FEM to investigate the SPDT process of silicon and silicon carbide, respectively. A few analytical studies (4–7%) have also been reported on SPDT process. Gradually, many researchers started using FEM to simulate micro-nano-level machining processes.

By designing a mathematical model in C++ and verifying the findings with finite element simulation carried out with ANSYS™, Shaini and Khera [118] examined the nonlinear creep behaviour of silicon nitride. Ji et al. [119] employed FEM-based numerical simulation of micro-grinding process of SiC ceramics to analyse the results prior to the grinding experiments. Further, Nian [120] and Duan et al. [121] numerically investigated the damage interference when scratching single crystal SiC with two cone-shaped diamond grits by combining the FEM and SPH methods. Figure 10 depicts the different FEM simulation models reported in the literature.

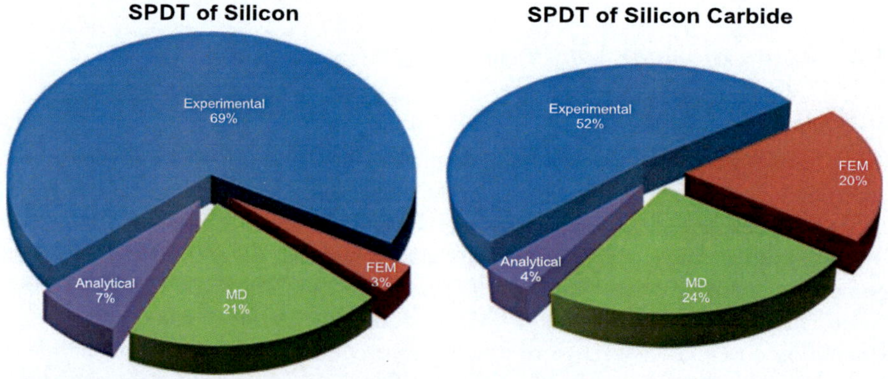

Fig. 9 Reported publications of different machining modes

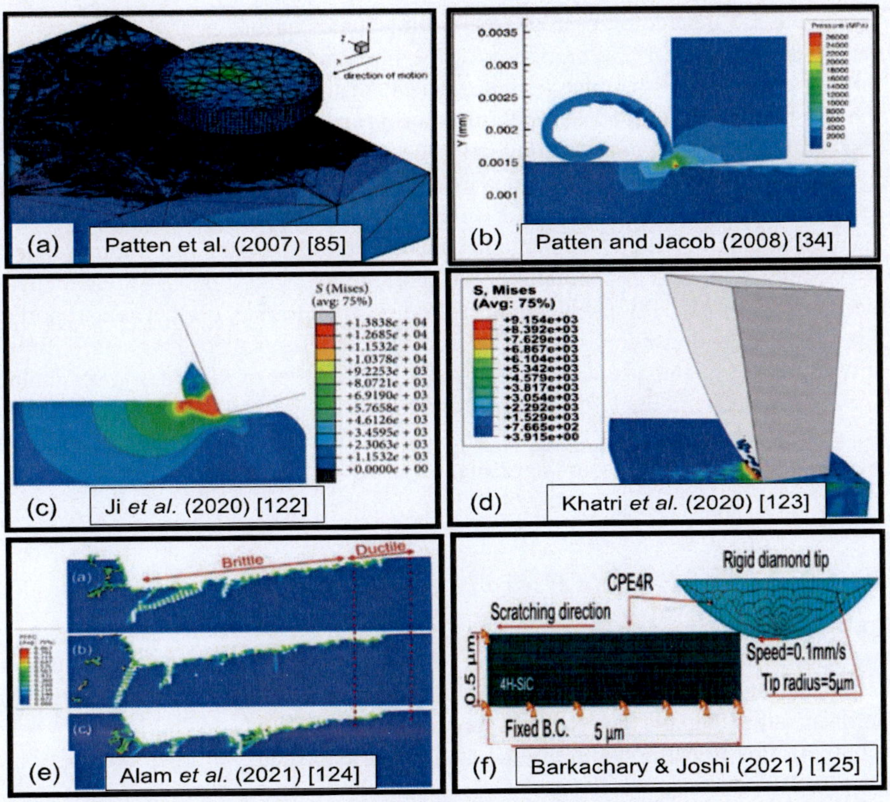

Fig. 10 FEM simulation models and results of SPDT **a** Patten et al. [85], **b** Patten and Jacob [34], **c** Ji et al. [122], **d** Khatri et al. [123], **e** Alam et al. [124] and **f** Barkachary and Joshi [125]

Using finite element modelling, Yan et al. [126] explored the role of hydrostatic pressure and temperature on the ductile machining of silicon. The authors found that high pressure, not the high temperature, is the main factor of ductile machining of silicon. Following the similar type of FEM approach to model SPDT of silicon, Yan et al. [76] affirmed that increasing the cutting-edge radius induces a reduction in uncut chip thickness and a commensurate rise in thrust force. Authors reported that if the cutting-edge radii are reduced lower than 200 nm leads the high temperature zone to relocate from the tool rake face to the tool flank face, turning the tool wear pattern from crater to flank wear. Figures 10a, b represent the work of Patten et al. [33, 85] and Patten and Jacob [34] in simulating SPDT of monocrystalline 6H–SiC using a Drucker–Prager (pressure sensitive) yield criteria in a commercially available FEM tool. Patten et al. [85] presented a three-dimensional (3D) scratching simulation of both Si and CVD-SiC with a stylus-radius diamond tip of 5 µm to determine the DBT depths. The authors attempted to investigate the DBT by analysing the machining forces influenced by rake angle and depth of cut. It was reported that only under ductile regime machining case did the cutting forces accord well with the experimental data. Ji et al. [122] carried out experiment and FEM simulation on silicon carbide to investigate the micro-grinding mechanism (Fig. 10c). The studied results showed that grinding depth is important during machining and a modest negative rake angle is advantageous for obtaining the optical surface on a SiC component. Khatri et al. [123] performed diamond machining of silicon by incorporating surface defects during machining to enhance performance of diamond machining (Fig. 10d). 3D FEM simulations were supplemented to support the experimental findings. Alam et al. [124] reported FEM simulation of diamond grooving on 3C–SiC to analyse the mechanism of DBT using Drucker–Prager model (Fig. 10e). It was reported that the DBT is dependent on depth of cut and independent of rake angle. Similarly, Barkachary and Joshi [125] reported a FEM model of nanometric scratching of SiC to understand the UPM process, and the results are validated with published experimental results (Fig. 10f). The reported experimental findings and the predicted machining forces showed good agreement.

3.3 *Smoothed Particle Hydrodynamics (SPH)*

As reported in Mir [127], Gingold and Monaghan pioneered the SPH method for astrophysical applications in 1977. SPH formulation is based on the interaction of particles rather than the mesh based approach in case of FEM simulation. As a result, it outperforms FEM modelling in the analysis of enormous deformation processes but falls short in the investigation of processes with tensile instability.

Mir et al. [128] conducted experiments and SPH simulation to uncover the underlying phenomena of DBT and diamond tool wear during SPDT machining of silicon (Fig. 11a). Liu et al. [132] performed a single grain scratching simulation to investigate the SiC grinding process using SPH approach. In this study, the material removal process, fracture initiation and propagation, surface roughness and scratching force were all thoroughly examined. Mir et al. [129] further employed SPH to perform numerical simulation of surface defect machining (SDM) of silicon, as illustrated in Fig. 11b. It was discovered that the SDM technique minimises the cutting resistance and energy of the material thereby lowering the diamond tool wear and surface roughness. Authors also found that when SDM is used, the von-Mises stress is somewhat less than when SDM is not used. Zhang and Zong [130] employed three types of tool-edge micro-defects, viz. blunt-edge, crescent-edge and flat-edge micro-defects in SPH simulation of SPDT of KDP crystal to analyse the material removal mechanism (Fig. 11c). Report revealed that the best tool for generating a smooth surface with the least surface roughness was found to have a chamfered edge. Klippel et al. [131] performed single grain diamond cutting simulation on silicon using SPH considering the temperature and hydrostatic pressure effect to obtain stress and temperature distribution, cutting forces and DBT behaviour (Fig. 11d). When simulation outputs are

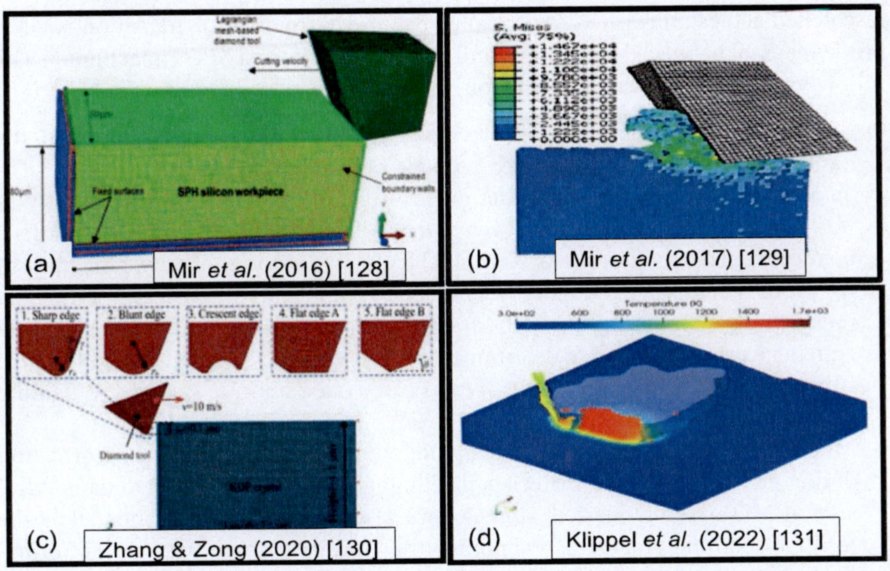

Fig. 11 SPH simulations model and results of SPDT process **a** Mir et al. [128], **b** Mir et al. [129], **c** Zhang and Zong [130] and **d** Klippel et al. [131]

compared to single grain scratching trials, it is discovered that cutting forces in simulation are somewhat greater. It was also discovered that the used Johnson–Cook flow stress model is unable of predicting DBT thickness alone.

4 Conclusions

Machining forces and surface roughness must be extensively examined in the UPM processes to determine product quality and process efficiency. In this regard, substantial research works on comprehending the cutting mechanism and chip generation during the machining of brittle and ductile materials have been reported. The challenging nanometric cutting process of brittle materials is studied using numerical simulation, which is considered as an easy, effective and cost-effective alternative tool. Simulation of SPDT using Molecular dynamics, finite element technique and smoothed particle hydrodynamics method have all received significant attention from researchers throughout the world. They concentrated on the investigation of DRM, DBT and critical depth of cut, process parameters, cutting forces, stress, strain, chips formation, shear angle, temperature, crystallographic effect, phase transformation, surface finish and tool wear during the UPM operation. An exhaustive literature review on experimental and numerical investigations on SPDT of brittle and ductile materials was conducted in the current work. The reported findings provide highly helpful information on SPDT of brittle materials in ductile mode. Many scholars researched and established various strategies for determining the transition zone of brittle material between brittle mode and ductile mode during SPDT machining. The following summaries were derived from the comprehensive literature study.

- SPDT is capable of machining any engineering material as it uses diamond, the hardest material as cutting tool to produce optical surface finish. However, it is an unit micro removal process that makes it very time-consuming and costly as it is associated with severe tool wear. Analysis of machining forces and surface quality of machined surface is vital for calculating power, economy, efficiency of the process and product quality of produced component. Extensive experimental inquiry, expensive in terms of both cost and time, is required to analyse and anticipate tool failure and to determine the optimal process parameters. Numerical simulation can be a novel alternative to carry out extensive trials for obtaining optimal process parameters.
- Much experimental work on understanding the cutting process and chip formation during ductile and brittle materials machining has been published to date. Also, several MD-based numerical studies have also been reported to understand the SPDT machining process by examining the machining force, stress, chip formation, phase change and temperature during the machining process. But, owing to atomic-level study domain, experimental confirmation of these findings is still challenging.
- The literature on numerical simulations with the finite element method is fairly scarce. The numerical studies of the SPDT process assist in understanding the mechanism of the process, such as the complex too-work interaction, material behaviour (transition) and many more. The machining forces, DBT thickness, chip form, and surface roughness may all be calculated using simulations.
- However, the numerical simulations are based on simplified assumption and reduced model of the actual machined work-tool geometry. Hence, incorporation of all the parameters including effect of uncontrollable parameters such as machine tool vibration, error in the setting of cutting tools, deformation of tool, workpiece fixtures and material inhomogeneity is still limits the use of numerical methods for completely replacing actual experimental study.
- Therefore, there is still a need of developing a better numerical model which is capable of incorporating all the process parameters that is affects the experimental process to completely mimic the actual SPDT process.

Acknowledgements This paper is an expanded version of an article entitled, "Ultra-precise single-point diamond turning process and its low-cost alternative methods" presented in "North East Research Conclave 2022" held at Indian Institute of Technology Guwahati, North Guwahati, India, during 20-22 May 2022. The work presented in the paper has not received any grants or support.

References

1. Davis GE, Roblee JW, Hedges AR (2009) Comparison of freeform manufacturing techniques in the production of monolithic lens arrays. In: Optical manufacturing and testing VIII (vol 7426). International Society for Optics and Photonics, p 742605
2. Davies MA, Evans CJ, Vohra RR, Bergner BC, Patterson SR (2003) Application of precision diamond machining to the manufacture of microphotonics components. In: Lithographic and micromachining techniques for optical component fabrication II (vol 5183). International Society for Optics and Photonics, pp 94–109
3. Ikawa N, Donaldson RR, Komanduri R, König W, McKeown PA, Moriwaki T, Stowers IF (1991) Ultraprecision metal cutting—the past, the present and the future. CIRP Ann Manuf Technol 40(2):587–594
4. Yu DP, San Wong Y, Hong GS (2011) A novel method for determination of the subsurface damage depth in diamond turning of brittle materials. Int J Mach Tools Manuf 51(12):918–927
5. Sumipro, https://www.sumipro.nl/technicaloptics/. Last Accessed on 12 June 2018
6. Chiu WM, Lee WB (1997) Development of ultra-precision machining technology
7. Goel S (2013) An atomistic investigation on the nanometric cutting mechanism of hard, brittle materials. Doctoral dissertation, Heriot-Watt University
8. Optics & Allied Engineering Pvt. Ltd., http://www.opticsindia.com/products/ir-optics-and-spdt/spdt-applications-areas/. Last Accessed on 10 Jan 2019
9. Fang FZ, Zhang XD, Weckenmann A, Zhang GX, Evans C (2013) Manufacturing and measurement of freeform optics. CIRP Ann 62(2):823–846
10. Krauskopf B (1984) Diamond turning: reflecting demands for precision. Manuf Eng 92(5):90–100
11. King RF, Tabor D (1954) The strength properties and frictional behaviour of brittle solids. Proc R Soc Lond A 223(1153):225–238
12. Leung TP, Lee WB, Lu XM (1998) Diamond turning of silicon substrates in ductile-regime. J Mater Process Technol 73(1–3):42–48
13. Patten JA, Gao W (2001) Extreme negative rake angle technique for single point diamond nano-cutting of silicon. Precis Eng 25(2):165–167
14. Yan J, Yoshino M, Kuriagawa T, Shirakashi T, Syoji K, Komanduri R (2001) On the ductile machining of silicon for micro electro-mechanical systems (MEMS), opto-electronic and optical applications. Mater Sci Eng, A 297(1–2):230–234
15. Yoshino M, Aoki T, Shirakashi T, Komanduri R (2001) Some experiments on the scratching of silicon: In situ scratching inside an SEM and scratching under high external hydrostatic pressures. Int J Mech Sci 43(2):335–347
16. Kong MC, Lee WB, Cheung CF, To S (2006) A study of materials swelling and recovery in single-point diamond turning of ductile materials. J Mater Process Technol 180(1–3):210–215
17. Mishra V, Garg H, Karar V, Khan GS (2019) Ultra-precision diamond turning process. In: Micro and nano machining of engineering materials. Springer, Cham, pp 65–97
18. Ding X, Rahman M (2012) A study of the performance of cutting polycrystalline Al 6061 T6 with single crystalline diamond micro-tools. Precis Eng 36(4):593–603
19. Kushendarsyah S, Sathyan S (2013) Orthogonal microcutting of thin workpieces. J Manuf Sci Eng 135(3):031004
20. Fang FZ, Huang KT, Gong H, Li ZJ (2014) Study on the optical reflection characteristics of surface micro-morphology generated by ultra-precision diamond turning. Opt Lasers Eng 62:46–56
21. Luo X, Tong Z, Liang Y (2014) Investigation of the shape transferability of nanoscale multi-tip diamond tools in the diamond turning of nanostructures. Appl Surf Sci 321:495–502
22. Mahajan KA, Sadaiah M, Gawande SH (2010) Experimental investigations of surface roughness on OFHC copper by diamond turning machine. Int J Eng Sci Technol 2(10):5215–5220
23. Zong WJ, Huang YH, Zhang YL, Sun T (2014) Conservation law of surface roughness in single point diamond turning. Int J Mach Tools Manuf 84:58–63

24. Zhou M, Zhang HJ, Chen SJ (2010) Study on diamond cutting of nonrationally symmetric microstructured surfaces with fast tool servo. Mater Manuf Process 25(6):488–494
25. Lin ZC, Huang JC, Jeng YR (2007) 3D nano-scale cutting model for nickel material. J Mater Process Technol 192:27–36
26. Wang H, To S (2011) A study on tool wear in ultra-precision diamond turning with finite element modelling. Int J Nanomanuf 7(5–6):500–516
27. Yu DP, Hong GS, San Wong Y (2012) Profile error compensation in fast tool servo diamond turning of micro-structured surfaces. Int J Mach Tools Manuf 52(1):13–23
28. Fang FZ (1998) Nano-turning of single crystal silicon. J Mater Process Technol 82(1–3):95–101
29. Yan J, Syoji K, Tamaki JI (2003) Some observations on the wear of diamond tools in ultra-precision cutting of single-crystal silicon. Wear 255(7–12):1380–1387
30. Fang FZ, Wu H, Zhou W, Hu XT (2007) A study on mechanism of nano-cutting single crystal silicon. J Mater Process Technol 184(1–3):407–410
31. Goel S, Luo X, Reuben RL (2013) Wear mechanism of diamond tools against single crystal silicon in single point diamond turning process. Tribol Int 57:272–281
32. Singh K, Vaishya RO, Singh H, Mishra V, Ramagopal S (2013) Investigation of tool life and surface roughness during single point diamond turning of silicon. Int J Sci Res 2(6):265–267
33. Patten J, Gao W, Yasuto K (2005) Ductile regime nanomachining of single-crystal silicon carbide. J Manuf Sci Eng 127(3):522–532
34. Patten JA, Jacob J (2008) Comparison between numerical simulations and experiments for single-point diamond turning of single-crystal silicon carbide. J Manuf Process 10(1):28–33
35. Ravindra D, Patten JA, Qu J (2009) Single point diamond turning effects on surface quality and subsurface damage in ceramics. In ASME 2009 international manufacturing science and engineering conference. American Society of Mechanical Engineers, pp 707–713
36. Luo X, Goel S, Reuben RL (2012) A quantitative assessment of nanometric machinability of major polytypes of single crystal silicon carbide. J Eur Ceram Soc 32(12):3423–3434
37. Goel S, Luo X, Comley P, Reuben RL, Cox A (2013) Brittle–ductile transition during diamond turning of single crystal silicon carbide. Int J Mach Tools Manuf 65:15–21
38. Goel S, Stukowski A, Luo X, Agrawal A, Reuben RL (2013) Anisotropy of single-crystal 3C–SiC during nanometric cutting. Modell Simul Mater Sci Eng 21(6):065004
39. Blackley WS, Scattergood RO (1989) Diamond turning of brittle materials. 5th Int Precision Eng Seminar 18(22):9
40. Patten JA (1996) High pressure phase transformation analysis and molecular dynamics simulations of single point diamond turning of Germanium
41. Yan J, Takahashi Y, TAMAKI JI, Kubo A, Kuriyagawa T, Sato Y (2006) Ultraprecision machining characteristics of poly-crystalline germanium. JSME Int J Series C Mech Syst, Mach Elements Manuf 49(1):63–69
42. Pawase P, Brahmankar PK, Pawade RS, Balasubramanium R (2014) Analysis of machining mechanism in diamond turning of germanium lenses. Procedia Mater Sci 5:2363–2368
43. Gupta S, Khatri N, Karar V, Dhami SS (2016) Investigation of surface roughness of single point diamond turned germanium substrate by coherence correlation interferometry and image processing. In: IOP conference series: materials science and engineering (vol 149, No 1). IOP Publishing, p 012032
44. Fang FZ, Liu XD, Lee LC (2003) Micro-machining of optical glasses—a review of diamond-cutting glasses. Sadhana 28(5):945–955
45. Fang FZ, Zhang GX (2004) An experimental study of optical glass machining. Int J Adv Manuf Technol 23(3–4):155–160
46. Bhagwat Vishal B, Manu R, Sreenath AM (2012) Ductile mode machining simulation of glass material using MATLAB. 2(4)
47. Bifano TG, Dow TA, Scattergood RO (1991) Ductile-regime grinding: a new technology for machining brittle materials. J Eng Ind 113(2):184–189
48. Anand PSP (2019) Significance of diamond as a cutting tool in ultra-precision machining process. In: Some aspects of diamonds in scientific research and high technology. IntechOpen

49. Lawn BR, Wilshaw R (1975) Indentation fracture: principles and applications. J Mater Sci 10(6):1049–1081
50. Lawn BR, Evans AG (1977) A model for crack initiation in elastic/plastic indentation fields. J Mater Sci 12(11):2195–2199
51. Lawn BR, Evans AG, Marshall DB (1980) Elastic/plastic indentation damage in ceramics: the median/radial crack system. J Am Ceram Soc 63(9–10):574–581
52. Marshall DB, Lawn BR (1985) Indentation of brittle materials. In: Microindentation techniques in materials science and engineering. ASTM International
53. Toh SB, McPherson R (1986) Fine scale abrasive wear of ceramics by a plastic cutting process. In: Almond EA, Brookes CA, Warren R (eds) Science of hard materials. Proceedings of 2nd international conference on science of hard materials held at rhodes 23–28 Sept 1984. Adam Hilger, Bristol, p 865
54. Moriwaki T, Shamoto E, Inoue K (1992) Ultraprecision ductile cutting of glass by applying ultrasonic vibration. CIRP Ann 41(1):141–144
55. Blackley WS, Scattergood RO (1994) Chip topography for ductile-regime machining of germanium. J Eng Ind 116(2):263–266
56. Fang FZ, Venkatesh VC (1998) Diamond cutting of silicon with nanometric finish. CIRP Ann 47(1):45–49
57. Huerta M, Malkin S (1976) Grinding of glass: the mechanics of the process. J Eng Ind 98(2):459–467
58. Moore MA, King FS (1980) Abrasive wear of brittle solids. Wear 60(1):123–140
59. Blake PN, Scattergood RO (1989) Ductile-regime turning of germanium and silicon
60. Blake PN, Scattergood RO (1990) Ductile-regime machining of germanium and silicon. J Am Ceram Soc 73(4):949–957
61. Yan J, Takahashi H, Gai X, Harada H, Tamaki JI, Kuriyagawa T (2006) Load effects on the phase transformation of single-crystal silicon during nanoindentation tests. Mater Sci Eng, A 423(1–2):19–23
62. Rao R, Bradby JE, Ruffell S, Williams JS (2007) Nanoindentation-induced phase transformation in crystalline silicon and relaxed amorphous silicon. Microelectron J 38(6–7):722–726
63. Goel S, Faisal NH, Luo X, Yan J, Agrawal A (2014) Nanoindentation of polysilicon and single crystal silicon: molecular dynamics simulation and experimental validation. J Phys D Appl Phys 47(27):275304
64. Goel S, Yan J, Luo X, Agrawal A (2014) Incipient plasticity in 4H–SiC during quasistatic nanoindentation. J Mech Behav Biomed Mater 34:330–337
65. Meng B, Zhang F, Li Z (2015) Deformation and removal characteristics in nanoscratching of 6H-SiC with Berkovich indenter. Mater Sci Semicond Process 31:160–165
66. Arif M, Xinquan Z, Rahman M, Kumar S (2013) A predictive model of the critical undeformed chip thickness for ductile–brittle transition in nano-machining of brittle materials. Int J Mach Tools Manuf 64:114–122
67. Xiao G, Ren M, To S (2018) A study of mechanics in brittle-ductile cutting mode transition. Micromachines 9(2):49
68. Yan J, Asami T, Harada H, Kuriyagawa T (2012) Crystallographic effect on subsurface damage formation in silicon microcutting. CIRP Ann Manuf Technol 61(1):131–134
69. Fang FZ, Zhang GX (2003) An experimental study of edge radius effect on cutting single crystal silicon. Int J Adv Manuf Technol 22(9–10):703–707
70. Wu H, Melkote SN (2012) Study of ductile-to-brittle transition in single grit diamond scribing of silicon: application to wire sawing of silicon wafers. J Eng Mater Technol 134(4)
71. Yin L, Vancoille EYJ, Lee LC, Liu YC, Huang H, Ramesh K (2003) High-precision low-damage grinding of polycrystalline SiC. In: Key engineering materials (vol 238). Trans Tech Publications, pp 59–64
72. Yuan YC, Li BZ, Zhou ZX, Zhang Q (2012) Study on the simulation model and characteristics of high-speed grinding for ceramics. In: Applied mechanics and materials (vol 138). Trans Tech Publications, pp 662–667

73. Zhu D, Yan S, Li B (2014) Single-grit modeling and simulation of crack initiation and propagation in SiC grinding using maximum undeformed chip thickness. Comput Mater Sci 92:13–21
74. Beltrão PA, Gee AE, Corbett J, Whatmore RW, Goat CA, Impey SA (1999) Single point diamond machining of ferroelectric materials. J Eur Ceram Soc 19(6–7):1325–1328
75. Ngoi BKA, Sreejith PS (2000) Ductile regime finish machining-a review. Int J Adv Manuf Technol 16(8):547–550
76. Yan J, Asami T, Harada H, Kuriyagawa T (2009) Fundamental investigation of subsurface damage in single crystalline silicon caused by diamond machining. Precis Eng 33(4):378–386
77. Yan J, Zhao H, Kuriyagawa T (2009) Effects of tool edge radius on ductile machining of silicon: an investigation by FEM. Semicond Sci Technol 24(7):075018
78. Lawn BR, Padture NP, Cait H, Guiberteau F (1994) Making ceramics "ductile." Science 263(5150):1114–1116
79. Nakasuji T, Kodera S, Hara S, Matsunaga H, Ikawa N, Shimada S (1990) Diamond turning of brittle materials for optical components. CIRP Ann Manuf Technol 39(1):89–92
80. Bridgman PW, Šimon I (1953) Effects of very high pressures on glass. J Appl Phys 24(4):405–413
81. Morris JC, Callahan DL, Kulik J, Patten JA, Scattergood RO (1995) Origins of the ductile regime in single-point diamond turning of semiconductors. J Am Ceram Soc 78(8):2015–2020
82. Shimada S, Ikawa N, Inamura T, Takezawa N, Ohmori H, Sata T (1995) Brittle-ductile transition phenomena in microindentation and micromachining. CIRP Ann Manuf Technol 44(1):523–526
83. Yan J, Syoji K, Kuriyagawa T, Suzuki H (2002) Ductile regime turning at large tool feed. J Mater Process Technol 121(2–3):363–372
84. Bhattacharya B, Patten JA, Jacob J (2006) Single point diamond turning of CVD coated silicon carbide. In: ASME 2006 international manufacturing science and engineering conference. American Society of Mechanical Engineers, pp 1153–1158
85. Patten JA, Jacob J, Bhattacharya B, Grevstad A, Fang N, Marsh ER (2007) Numerical simulations and cutting experiments on single point diamond machining of semiconductors and ceramics. Semiconductor machining at the micro-nano scale. Transw Res Netw Kerala India 2:1–36
86. Ravindra D, Patten J (2007) Determining the ductile to brittle transition (DBT) of a single-crystal 4H-SiC wafer by performing nanometric cutting. In: Proceedings of ISAAT 2007 on precision grinding and abrasive technology at SME international grinding conference, pp 26–28
87. Kovalchenko AM (2013) Studies of the ductile mode of cutting brittle materials (a review). J Superhard Mater 35(5):259–276
88. Hatefi S, Abou-El-Hossein K (2020) Review of single-point diamond turning process in terms of ultra-precision optical surface roughness. Int J Adv Manuf Technol 106(5):2167–2187
89. Alder BJ, Wainwright T (1957) Phase transition for a hard sphere system. J Chem Phys 27(5):1208–1209
90. Belak JF, Stowers IF (1990) A molecular dynamics model of the orthogonal cutting process (No. UCRL-102697; CONF-9009241--1). Lawrence Livermore National Lab., CA (USA)
91. Liu Y, Li B, Kong L (2018) Atomistic insights on the nanoscale single grain scratching mechanism of silicon carbide ceramic based on molecular dynamics simulation. AIP Adv 8(3):035109
92. Liu Y, Li B, Kong L (2018) A molecular dynamics investigation into nanoscale scratching mechanism of polycrystalline silicon carbide. Comput Mater Sci 148:76–86
93. Liu Y, Li B, Kong L (2018) Molecular dynamics simulation of silicon carbide nanoscale material removal behavior. Ceram Int 44(10):11910–11913
94. Wang J, Chi H, Lv M, Liu X, Li Y, Zhao Y (2020) Effect of silicon carbide hard particles scratch on the diamond cutting tools groove wear. Proc Inst Mech Eng C J Mech Eng Sci 234(10):2053–2063

95. Shimada S, Ikawa N (1992) Molecular dynamics analysis as compared with experimental results of micromachining. CIRP Ann 41(1):117–120
96. Inamura T, Shimada S, Takezawa N, Nakahara N (1997) Brittle/ductile transition phenomena observed in computer simulations of machining defect-free monocrystalline silicon. CIRP Ann 46(1):31–34
97. Komanduri R, Chandrasekaran N, Raff LM (2000) MD Simulation of nanometric cutting of single crystal aluminum–effect of crystal orientation and direction of cutting. Wear 242(1–2):60–88
98. Komanduri R, Chandrasekaran N, Raff LM (2001) Molecular dynamics simulation of the nanometric cutting of silicon. Philos Mag B 81(12):1989–2019
99. Cheng K, Luo X, Ward R, Holt R (2003) Modeling and simulation of the tool wear in nanometric cutting. Wear 255(7–12):1427–1432
100. Cai MB, Li XP, Rahman M (2007) Characteristics of "dynamic hard particles" in nanoscale ductile mode cutting of monocrystalline silicon with diamond tools in relation to tool groove wear. Wear 263(7–12):1459–1466
101. Komanduri R, Raff LM (2001) A review on the molecular dynamics simulation of machining at the atomic scale. Proc Inst Mech Eng, Part B: J Eng Manuf 215(12):1639–1672
102. Oluwajobi A (2012) Molecular dynamics simulation of nanoscale machining. In: Molecular dynamics-studies of synthetic and biological macromolecules. InTech
103. Goel S, Luo X, Agrawal A, Reuben RL (2015) Diamond machining of silicon: a review of advances in molecular dynamics simulation. Int J Mach Tools Manuf 88:131–164
104. Guo X, Li Q, Liu T, Kang R, Jin Z, Guo D (2017) Advances in molecular dynamics simulation of ultra-precision machining of hard and brittle materials. Front Mech Eng 12(1):89–98
105. Abdulkadir LN, Abou-El-Hossein K, Jumare AI, Liman MM, Olaniyan TA, Odedeyi PB (2018) Review of molecular dynamics/experimental study of diamond-silicon behavior in nanoscale machining. Int J Adv Manuf Technol 1–55
106. Goel S, Luo X, Reuben RL, Rashid WB (2011) Atomistic aspects of ductile responses of cubic silicon carbide during nanometric cutting. Nanoscale Res Lett 6(1):589
107. Goel S, Luo X, Reuben RL (2012) Molecular dynamics simulation model for the quantitative assessment of tool wear during single point diamond turning of cubic silicon carbide. Comput Mater Sci 51(1):402–408
108. Goel S, Luo X, Reuben RL, Pen H (2012) Influence of temperature and crystal orientation on tool wear during single point diamond turning of silicon. Wear 284:65–72
109. Goel S, Kovalchenko A, Stukowski A, Cross G (2016) Influence of microstructure on the cutting behaviour of silicon. Acta Mater 105:464–478
110. Li XP, He T, Rahman M (2005) Tool wear characteristics and their effects on nanoscale ductile mode cutting of silicon wafer. Wear 259(7–12):1207–1214
111. Liu C, He W, Chu J, Zhang J, Chen X, Xiao J, Xu J (2021) Molecular dynamics simulation on cutting mechanism in the hybrid machining process of single-crystal silicon. Nanoscale Res Lett 16(1):1–15
112. Goel S, Chavoshi SZ, Murphy A (2017) Molecular dynamics simulation (MDS) to study nanoscale machining processes. In: Jain VK (ed), Nanofinishing science and technology: basic and advanced finishing and polishing processes (micro and nanomanufacturing series). Taylor and Francis
113. Davim JP, Maranhao C, Jackson MJ, Cabral G, Gracio J (2008) FEM analysis in high speed machining of aluminium alloy (Al7075-0) using polycrystalline diamond (PCD) and cemented carbide (K10) cutting tools. Int J Adv Manuf Technol 39(11–12):1093–1100
114. Davim JP, Reis P, Maranhao C, Jackson MJ, Cabral G, Gracio J (2009) Finite element simulation and experimental analysis of orthogonal cutting of an aluminium alloy using polycrystalline diamond tools. Int J Mater Prod Technol 37(1–2):46–59
115. Jagadesh T, Samuel GL (2015) Mechanistic and finite element model for prediction of cutting forces during micro-turning of titanium alloy. Mach Sci Technol 19(4):593–629
116. Jagadesh T, Samuel GL (2017) Finite element simulations of micro turning of Ti–6Al–4V using PCD and coated carbide tools. J Inst Eng (India): Series C 98(1):5–15

117. Markopoulos AP (2012) Finite element method in machining processes. Springer Science & Business Media
118. Saini JS, Khera S (2016) Non-linear creep analysis of ceramic specimen using finite element method. J Inst Eng (India): Series C 97(3):417–430.
119. Ji S, Liu L, Zhao J, Sun C (2015) Finite element analysis and simulation about microgrinding of SiC. Journal of Nanomaterials
120. Nian D (2020) Effects of depth of cutting on damage interferences during double scratching on single crystal SiC. Curr Comput-Aided Drug Des 10(6):519
121. Duan N, Yu Y, Shi W, Xiao Q, Liu Q (2021) Investigation on diamond damaged process during a single-scratch of single crystal silicon carbide. Wear 486:204099
122. Ji S, Liu L, Zhao J, Sun C (2015) Finite element analysis and simulation about microgrinding of SiC. J Nanomater 575398:1–9
123. Khatri N, Barkachary BM, Muneeswaran B, Al-Sayegh R, Luo X, Goel S (2020) Surface defects incorporated diamond machining of silicon. Int J Extreme Manuf 2(4):045102
124. Alam M, Zhao L, Vajragupta N, Zhang J, Hartmaier A (2021) Finite Element modeling of brittle and ductile modes in cutting of 3C-SiC. Curr Comput-Aided Drug Des 11(11):1286
125. Barkachary BM, Joshi SN (2022) Numerical computation and analysis of cutting forces during nanometric scratching of silicon carbide. J Inst Eng (India): Series C 103(1):53–62
126. Yan J, Zhao H, Kuriyagawa T (2007) finite element analysis of the hydrostatic pressure and temperature in ductile machining of silicon. Adv Abrasive Technol 10:185
127. Mir A (2016) Investigation of cutting mechanics in single point diamond turning of silicon. Doctoral dissertation, University of Strathclyde
128. Mir A, Luo X, Sun J (2016) The investigation of influence of tool wear on ductile to brittle transition in single point diamond turning of silicon. Wear 364:233–243
129. Mir A, Luo X, Siddiq A (2017) Smooth particle hydrodynamics study of surface defect machining for diamond turning of silicon. Int J Adv Manuf Technol 88(9–12):2461–2476
130. Zhang S, Zong W (2020) Micro defects on diamond tool cutting edge affecting the ductile-mode machining of KDP crystal. Micromachines 11(12):1102
131. Klippel H, Süssmaier S, Röthlin M, Afrasiabi M, Pala U, Wegener K (2021) Simulation of the ductile machining mode of silicon. Int J Adv Manuf Technol 115(5):1565–1578
132. Liu Y, Li B, Wu C, Kong L, Zheng Y (2018) Smoothed particle hydrodynamics simulation and experimental analysis of SiC ceramic grinding mechanism. Ceram Int 44(11):12194–12203

Fabrication of a Cost-Effective Bi-porous Composite Wick for Loop Heat Pipes

Toni Kumari, Chandan Nashine, and Manmohan Pandey

Nomenclature

LHP	Loop heat pipe
UTM	Universal testing machine
HRB	Hardness Rockwell B
CC	Compensation chamber
m_d	Dry weight of wick
m_a	Saturated weight of wick in the air
m_w	Saturated weight of wick in the water
V_w	Volume of sintered wick
ε	Porosity of wick
K	Permeability of wick
ρ_m	Density of material

1 Introduction

Loop heat pipe (LHP) is a two-phase passive heat transport device that transmits heat from the evaporator to the condenser using the evaporation and the condensation of

T. Kumari (✉) · C. Nashine · M. Pandey
Department of Mechanical Engineering, Indian Institute of Technology Guwahati, Guwahati, Assam, India
e-mail: ktoni@iitg.ac.in

C. Nashine
e-mail: nashine18@iitg.ac.in

M. Pandey
e-mail: manmohan@iitg.ac.in

S. N. Joshi et al. (eds.), *Low Cost Manufacturing Technologies*,
https://doi.org/10.1007/978-981-19-8452-5_9

working fluid [1]. It utilizes the latent heat of working fluid to transfer heat across a long distance with minimal temperature difference. Russian scientist Maydanik put up the first LHP in the 1970s for use in aircraft applications. LHPs are currently widely employed in numerous cooling applications, such as the cooling of electronic devices and thermal control in space systems. As the LHP is a passive device, supported by a capillary medium in the evaporator, it is less dependent on gravity, does not require any external power input and has no moving parts. Therefore, LHP is a very reliable, auto-control device and needs less maintenance, which makes it suitable for remote applications as well. The LHP has some additional benefits compare to normal heat pipes [2, 3], for example, separate lines for liquid and vapour, which reduces the counter-flow pressure drop and entrainment problem. Only the evaporator part contains the capillary wick which also reduces the pressure drop and therefore increases heat transport capacity.

Figure 1 is a schematic diagram of mLHP. It consists of an evaporator, a condenser, a compensation chamber, a vapour line and a liquid line. The evaporator incorporates capillary structures (wick), while the other components are simply smooth tubes. The heat source provides heat input to the saddle, which is attached to the evaporator. A part of the heat input is used to evaporate the working fluid from the primary wick's outer surface. While the remaining part of the heat input is transported back to the CC because of thermal conduction of the evaporator wall and wick and called heat leak. Heat leak is not desirable [4], because it increases the saturation temperature, increases the evaporator wall temperature, reduces the evaporation rate and makes it difficult to start up. The liquid is converted to vapour in the vapour grooves, and the menisci developed in the evaporator wick creates pumping force to push the generated vapour to condenser through the vapour line; it rejected heat in condenser. The liquid that has condensed is forced back through the liquid line to the evaporator after passing through the compensation chamber.

The capillary wick is an important component of the device, which provides the pumping force to the working fluid for the completion of the loop. It helps in the thermal and hydraulic locking between system components. Fabrication of wick is expensive and complex as it needs to control certain parameters like pore size, permeability, porosity and effective thermal conductivity. High thermal conductivity wick material has high heat transport capacity but also has high heat leak. Though low thermal conductivity wick material abates the heat leak problem [5, 6], it reduces the heat transport capacity [7]. To utilize the advantage of both high and low thermal conductivity wick materials, a composite wick, made of low and high thermal conductivity material is introduced [8, 9]. The compensation chamber side of the wick has a low thermal conductivity layer, and the vapour channel has a high thermal conductivity layer to reduce heat leak and improve evaporation rate respectively and startup performance [10]. Ji et al. [11] developed a three-layer composite wick. Studies showed that compared to the mono-porous wick, the composite wick has lower heat leakage. Xu et al. [12] introduced a double-layer copper composite wick with a thickness of 2 mm in the liquid transportation zone and 3 mm thickness in the evaporation zone. They found that the optimum particle sizes for liquid movement and evaporation in the wick.

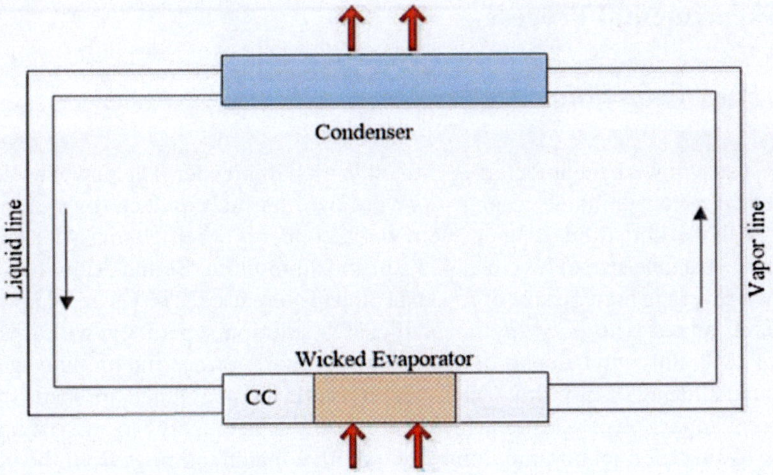

Fig. 1 Schematic diagram of mLHP

Similar to thermal conductivity, pore size also has a contradictory effect on wick characteristics [13–15]. Small pore size creates large capillary force, but it also develops resistance for fluid flow. Large pore size helps in fluid flow but reduces the sample strength and capillary force. To overcome the above problem bi-porous wick was introduced. Liu et al. [16] did an experimental investigation on a novel design of flat LHP. They considered bi-porous porous media and stainless steel mesh as primary wicks and secondary wicks, respectively. Wu et al. [17] fabricated a complex wick. The outer layer of the wick was bi-porous and had large pores size, which helps in vapour escape. The inner layer of the wick was mono-porous and had small pores size, increasing the strength and pumping force. The complex wick enhanced the maximum heat load by 67% compared to the mono-porous wick.

The literature review shows that there are very limited studies on the fabrication and characterization of composite bi-porous wicks. Although wicks are the heart of loop heat pipes, manufacturing them is complex and requires a precise balance of pore size, porosity, and permeability. Hence, to improve the performance of loop heat pipes, it is important to implement and better understand the characteristics of composite bi-porous wicks. To overcome the above challenges, the current work deals with the objective of the development of a cost-effective technique for the fabrication and characterization of a composite wick. In this paper, the details of the experimental setup are outlined and the characterization studies are carried out on bi-porous copper–aluminium composite wick for loop heat pipe.

2 Experimental Process

2.1 *Wick Fabrication Methodology*

The first step in wick manufacturing is the selection of powder. The powder selection is based on certain parameters such as powder size, thermal conductivity and compatibility with working fluid. In the present work, copper powder (Sigma-Aldrich) with an average particle size of less than 425 μm and aluminium (Sigma-Aldrich) powder having an average particle size of less than 10 μm were used. The Cu and Al powders were well mixed with polyvinyl alcohol (PVA) solution separately, with a volume ratio of 20%. Polyvinyl alcohol (PVA) works as a pore former and binding agent for the manufacturing of bi-porous wick. The reusable die and punch are made of mild steel for compaction of powder, which is cost-effective compared to an existing techniques like metal injection moulding and additive manufacturing. Both the powder was filled in mould one by one to fabricate a composite layer as shown in Fig. 2. The above arrangement was kept in UTM for compaction at 150 KN force for about 15 min. The compacted sample (green sample) is then sintered in a muffle furnace at 673 K for about 90 min. The PVA present in the sample evaporates at a higher temperature and creates pores. The sintered sample was allowed to cool down gradually for 10 h in the furnace. Figure 3 shows the fabricated bi-porous composite wick. The manufactured sample is 3 mm thick and has a diameter of 50 mm.

2.2 *Wick Parameters Testing*

2.2.1 Porosity

The porosity of sintered wick can be measured by two methods: (a) Archimedes method (density method) and (b) geometry method. In Archimedes's method, three independent weights are measured. First, dry weight (m_{d}) of the wick was measured using the electronic weighing machine. Further, the weight of saturated wick in the air (m_{a}) and weight of wick in the water (m_{w}) were measured, respectively. For the saturation of wick, the sample was kept in liquid (water) at 80 °C for about 7 h. Now porosity of wick can be calculated as:

$$\varepsilon = \frac{m_{\mathrm{a}} - m_{\mathrm{d}}}{m_{\mathrm{a}} - m_{\mathrm{w}}} \tag{1}$$

In the geometry method [18], the first volume of sintered wick (V_{w}) is measured by the geometry method and then the weight of the dry wick (m_{d}) is measured by an electronic weighing machine. Now, porosity of the wick can be calculated from the following relation:

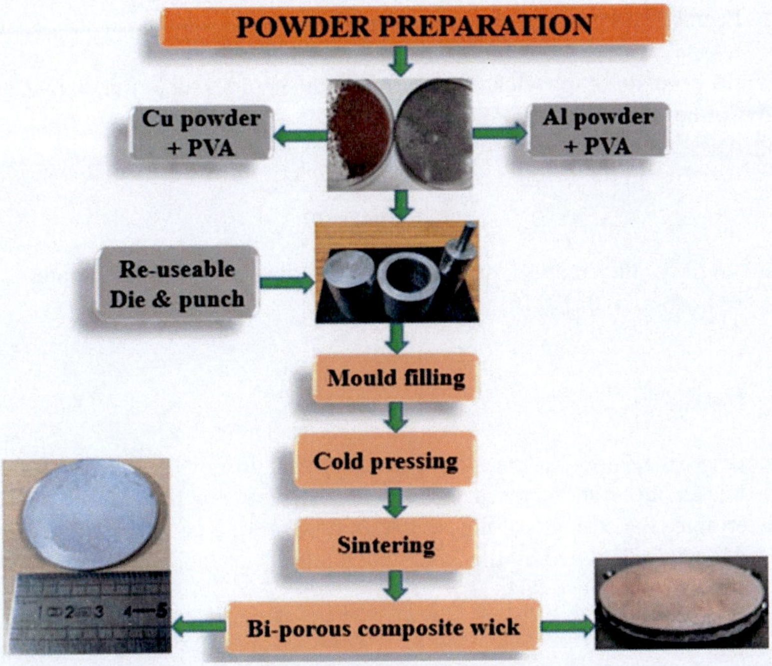

Fig. 2 Flowchart for fabrication process of bi-porous composite wick

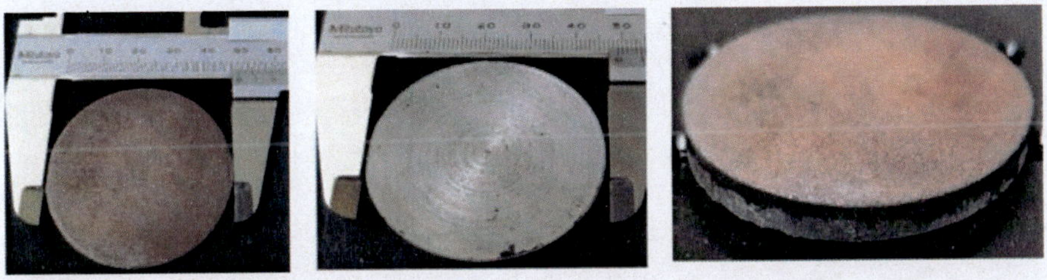

Fig. 3 Image of fabricated bi-porous composite wick **a** top side **b** bottom side **c** front view

$$\varepsilon = \frac{V_{\mathrm{w}}\rho_{\mathrm{m}} - m_{\mathrm{d}}}{V_{\mathrm{w}}\rho_{\mathrm{m}}} \quad (2)$$

The porosity of the wick was calculated at about 42%. It depends on particle shape, particle size and distribution of particles. In the case of the bi-porous wick, porosity also depends on pore former size, shape, type, and quantity of pore former (PVA). Porosity has negative relation with sintering temperature and sintering time. Therefore, if the sintering temperature is less porosity will be high and vice versa, the same correlation exists with sintering time. Due to pore former, a large pore size creates, which helps in easy escaping of vapour. But higher porosity will decrease the strength of the wick.

2.2.2 Permeability

If we know porosity of the wick, then permeability could be determined by Carman–Kozeny formula [18]:

$$k = \frac{d^2 \varepsilon^3}{180(1 - \varepsilon)^2} \tag{3}$$

Wick permeability reduces by increasing sintering time and temperature. It was calculated at about 0.0312 μm^2.

2.2.3 Hardness

The hardness measures the resistance to an indentation or localized deformation. Hardness is an important factor in deciding the machining method for the sample and also determines the strength of the sample in full working conditions. The hardness of the wick was calculated by the Rockwell hardness test and was found to be 35.6 HRB (Rockwell hardness scale B).

2.2.4 Pore Size

Scanning electron microscope (SEM) was used to study the pore size and structures of the wick sample. Figures 4a, b and 5a show the SEM morphology of the wick sample at a magnification of 1000, 5000 and 2000 times, respectively. In Fig. 5b, the pore size range varies from 70 to 1160 nm. The average pore diameter of the sintered wick is found to be 378 nm.

Fig. 4 Morphology of the sintered wick

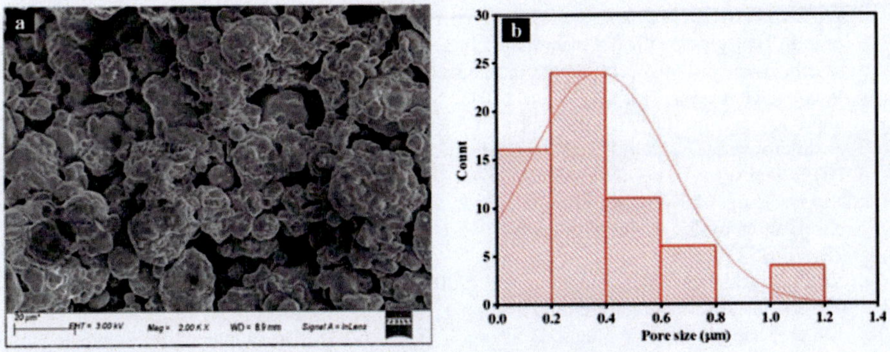

Fig. 5 **a** Morphology of the wick, **b** corresponding pore size distribution

3 Conclusion

In the current work, a disc-type bi-porous composite wick was successfully manufactured from copper and aluminium powder. The manufactured wick has the desired properties of pore size, porosity and permeability. The fabricated wick has the diameter of 50 mm and the thickness of 3 mm.

1. Implementation of re-useable die-punch made of mild steel for compaction is expected to be cost-effective in mass production of the wick.
2. Porosity and permeability of the wick was found to be 42% and 0.0312 μm^2.
3. The pore size of wick vary from 70 to 1160 nm. Maximum pores of the wick fall under the size range of 70–290 nm.
4. The fabricated bi-porous composite wick will help in reducing heat leak and is expected to improve the evaporation rate and start up performance of the LHP.

Acknowledgements The characterization experiments were conducted at the Central Instruments Facility of IIT Guwahati.

References

1. Ahmed S, Nashine C, Pandey M (2022) Thermal management at microscale level: detailed study on the development of a micro loop heat pipe. Micro Nano Eng:100150
2. Nashine C, Pandey M (2021) Thermo-hydraulic modelling of miniature loop heat pipes. In: ISHMT digital library. Begel House Inc
3. Ahmed S, Pandey M, Kawaji M (2022) Loop heat pipe design: an evaluation of recent research on the selection of evaporator, wick, and working fluid. J Therm Sci Eng Appl 14(7):070801
4. Joung W, Yu T, Lee J (2008) Experimental study on the loop heat pipe with a planar bifacial wick structure. Int J Heat Mass Transf 51(7–8):1573–1581

5. Wu S-C, Peng J-C, Lai S-R, Yeh C-C, Chen Y-M (2009) Investigation of the effect of heat leak in loop heat pipes with flat evaporator. In: 2009 4th International microsystems, packaging, assembly and circuits technology conference. IEEE, pp 348–351
6. Wang S, Zhang W, Zhang X, Chen J (2011) Study on start-up characteristics of loop heat pipe under low-power. Int J Heat Mass Transf 54(4):1002–1007
7. Nishikawara M, Nagano H (2014) Parametric experiments on a miniature loop heat pipe with PTFE wicks. Int J Therm Sci 85:29–39
8. Xu J, Zhang L, Hong X, Zhong J, Xuan J (2014) Experimental investigation and visual observation of loop heat pipes with two-layer composite wicks. Int J Heat Mass Transf 72:378–387
9. Singh R, Akbarzadeh A, Mochizuki M (2009) Effect of wick characteristics on the thermal performance of the miniature loop heat pipe. J Heat Transf 131(8)
10. Xin G, Zhang P, Chen Y, Cheng L, Huang T, Yin H (2018) Development of composite wicks having different thermal conductivities for loop heat pipes. Appl Therm Eng 136:229–236
11. Ji X, Wang Y, Jinliang X, Huang Y (2017) Experimental study of heat transfer and start-up of loop heat pipe with multiscale porous wicks. Appl Therm Eng 117:782–798
12. Xu J, Wang D, Zhuohuan H, Zhang L, Ye L, Zhou Y (2020) Effect of the working fluid transportation in the copper composite wick on the evaporation efficiency of a flat loop heat pipe. Appl Therm Eng 178:115515
13. Yeh C-C, Chen C-N, Chen Y-M (2009) Heat transfer analysis of a loop heat pipe with biporous wicks. Int J Heat Mass Transf 52(19–20):4426–4434
14. Li H, Liu ZC, Chen BB, Liu W, Li C, Yang J (2012) Development of biporous wicks for flat-plate loop heat pipe. Exp Therm Fluid Sci 37:91–97
15. Mathews AJ, Ranjan S, Inbaoli A, Sujith Kumar CS, Jayaraj S (2021) Optimization of the sintering parameters of a biporous copper-nickel composite wick for loop heat pipes. Mater Today: Proc 46:9297–9302
16. Liu ZC, Li H, Chen BB, Yang JG, Liu W (2012) Operational characteristics of flat type loop heat pipe with biporous wick. Int J Therm Sci 58:180–185
17. Wu SC, Lin FC, Chen SH, Yeh CC, Wang SK, Chen YM (2011) Manufacturing of the complex wick with double-layer bi-porous structure and applying to a loop heat. In: 10th International heat pipe symposium, pp 104–108
18. Dullien FAL (1991) Porous media: fluid transport and pore structure. Academic Press, New York

Development and Characterisation of Bi-porous Metallic Wick for Loop Heat Pipes

Chandan Nashine, Nadaf Arman Mohaddin, Rohit Kumar, Sandip Kumar Sarma, and Manmohan Pandey

1 Introduction

Loop heat pipe (LHP) is a capillary force-driven passive two-phase heat transfer device which has emerged as a solution to meet the acute demand for thermal management in space and terrestrial applications [1]. The LHP is a closed-system loop which utilises the phase-change mechanism for heat transfer to a long distance [2]. Figure 1 shows the schematic of LHP. The device consists of several components like an evaporator, compensation chamber, condenser and transport lines. The evaporator consists of a porous wick, which is the main component of the LHP as it provides the capillary driving force and the evaporation area. The working fluid evaporates when the heat load is applied to the evaporator. Further, it passes from the vapour line to the condenser, condensing and recirculating using the capillary force in the porous wick. So, the primary factor that decides the performance of the LHP is the maximum capillary pressure rise developed in the wick, which dictates the device's heat transport capacity [3]. Capillary wick is the heart of the LHP, as it organises heat and mass transfer in the evaporator and helps in the circulation of the working fluid around the loop [2, 4]. The capillary driving force developed in the porous wick should be

C. Nashine (✉) · N. A. Mohaddin · R. Kumar · S. K. Sarma · M. Pandey
Department of Mechanical Engineering, Indian Institute of Technology Guwahati, Guwahati, Assam, India
e-mail: nashine18@iitg.ac.in

N. A. Mohaddin
e-mail: a.nadaf@iitg.ac.in

R. Kumar
e-mail: rohit18e@iitg.ac.in

S. K. Sarma
e-mail: s.sandip@iitg.ac.in

M. Pandey
e-mail: manmohan@iitg.ac.in

S. N. Joshi et al. (eds.), *Low Cost Manufacturing Technologies*,
https://doi.org/10.1007/978-981-19-8452-5_10

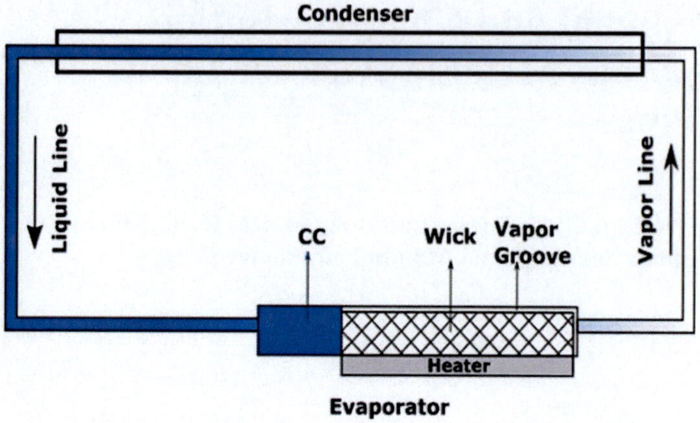

Fig. 1 Schematic of loop heat pipe

sufficiently high enough to overcome the total pressure drop inside the loop. High capillary pressure also ensures sufficient liquid feedback to prevent the dry-out condition in the evaporator. The porous capillary wick is generally fabricated using cold press sintering, additive manufacturing, metal injection moulding and 3D printing. The most critical parameter of porous wick includes the low hydraulic resistance to ensure a high flow rate and minimised effective pore radius to ensure high capillary pressure [5, 6]. Some studies related to the manufacturing of porous capillary wick are present in the literature. Their studies primarily emphasise the fabrication and characterisation of metallic and non-metallic wicks. Singh et al. [7] studied the properties of copper and nickel wicks used in the miniature loop heat pipe. In their later work [8], the authors presented the test procedure to measure wick properties and detail characterisation. Both metallic (copper) and plastic (polyethene) wicks were fabricated and studied. They found that wick having higher porosity, permeability and lower pore radius showed higher heat transfer characteristics.

Prasad et al. [9] performed the experimental measurement of pore size, porosity and thermal conductivity for a flat disc-shaped porous nickel wick. Wu et al. [10] presented the implementation of sintered polytetrafluoroethylene (PTFE) wick for LHP. They found that the pore size of the PTFE wick was about 300–500 μm. Deng et al. [11] studied the characterisation of four types of sintered wicks and compared the results and found that the copper wicks had higher capillary rise and permeability than nickel wicks. Seidel [12] investigated the values of permeability as a function of pore radius. Choi et al. [13] studied the low-temperature sintering of copper bi-porous wicks with improved capillary pressure. Samanta et al. [6] investigated the physical characteristics of porous wick fabricated using metal injection moulding. They found that the ideal temperature for sintering was about 60 min at 900 °C. Choi et al. [14] performed an experimental investigation on sintered porous wick for miniature loop heat pipe applications. They devised experimental modules to measure maximum capillary pressure and used the laser flash method for thermal conductivity measurement. Anvesh et al. [15] developed the porous nickel wick and performed the experimental characterisation. They outlined the details of the experimental setup used to characterise porous wick for use in LHP.

The above review of the literature shows that the difficulties associated with fabricating wick are the attainment of desired characteristics like porosity, pore radius and permeability. Although wicks are vital in LHP, manufacturing them is complicated and requires an optimum balance of critical elements like porosity, permeability, roundness, pore diameter, pore structure and capillary pumping pressure. The major production cost in the manufacturing of LHP is spent mainly on the fabrication of wick. In a wick, tiny pores should be employed for high capillary suction, but on the other hand, small pores will trap vapour and prevent rewetting. The wicks with single pore characterisation exhibit poor compromise between high capillary force and high vapour permeability. Therefore, to improve the performance of LHPs, it is essential to implement and better understand the characteristics of bi-porous wicks.

Therefore, aiming to address the issues mentioned above, the current work deals with developing and characterising low-cost bi-porous copper wicks. The main objective of the present work is to fabricate cost-effective bi-porous wicks for loop heat pipes to be used in the thermal management of space applications. Cylindrical and flat bi-porous copper wicks of different sizes are successfully fabricated for LHP with cylindrical and flat-shaped evaporators. The current work describes the particulars of the experimental setup and presents the characterisation studies on the copper bi-porous sintered wick to be used in loop heat pipes.

2 Experimental Methodology

In the current work, bi-porous metallic wicks of different sizes and shapes are fabricated using cold press sintering, as shown in Fig. 2. The flat wicks are fabricated with a disc shape of 49 mm in diameter and 5 mm thick. Cylindrical wicks are fabricated for different lengths and diameters as shown in Fig. 2a.

Copper powder (average particle size < 10 microns) was mixed well with polyvinyl alcohol (PVA) solution, which acts as a pore former and binding agent for bi-porous metallic wicks. Further, the fabrication of a reusable die and punch is done for the desired shape and size of the wick. The prepared mixture is then compacted using a die and punch set in the universal testing machine. The compaction is done at a force of 12 KN for 5 min.

The prepared sample is then sintered at three different temperatures and durations. Among these, the best sintering condition was identified as 650 °C and 90 min based on porosity achieved. The presence of pore former mixed into the copper powder ensures that the wick possess both fine pores and large interconnecting pores.

Figure 3 shows the details of the processes involved in the fabrication of copper bi-porous wicks. An isothermal muffle furnace is used for sintering. The sintering is performed in the presence of a rough vacuum of 0.001 Torr to prevent the sample from oxidation.

Fig. 2 Bi-porous wicks fabricated for loop heat pipes. **a** Cylindrical-shaped wicks. **b** A flat disc-shaped wick

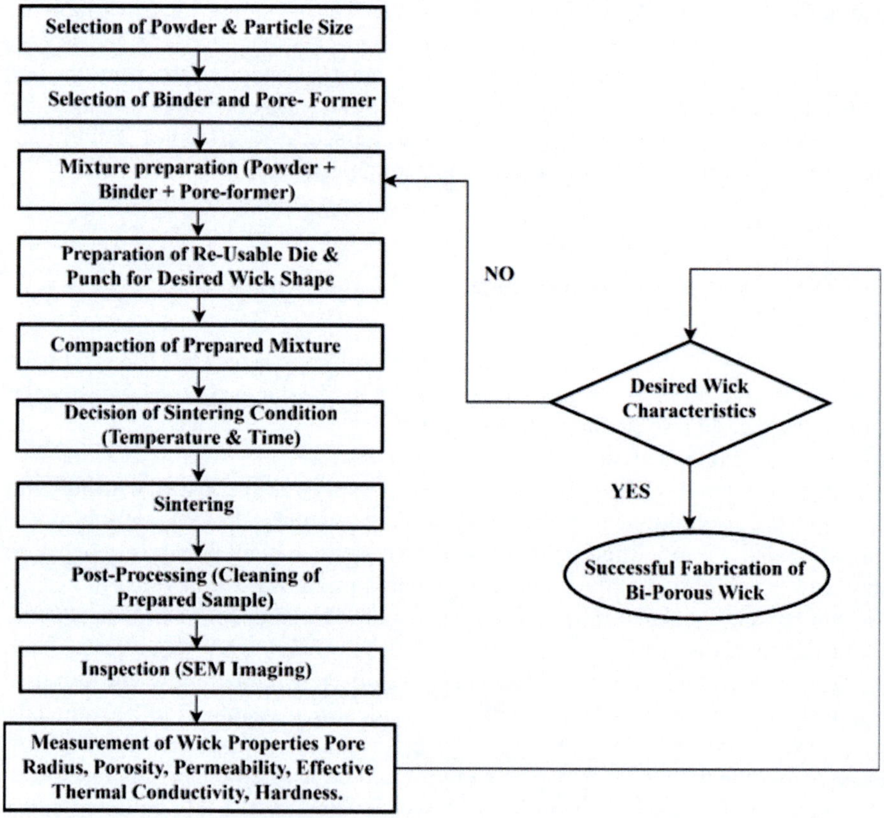

Fig. 3 Flowchart of the processes involved in the fabrication of bi-porous wick

3 Characterisation of Wick

3.1 Porosity

The porosity of any porous material is the ratio of the porous volume of the material to its total volume. In the current work, density method is adopted for the calculation of porosity [16]. Using density method, the porosity of the capillary wick can be estimated using the following expression:

$$\varepsilon = 1 - \frac{\rho_{\mathrm{w}}}{\rho_{\mathrm{cu}}} \quad (1)$$

Here, ε represents the porosity of the wick. ρ_{cu} and ρ_{w} represent the density of copper and the density of porous copper wick, respectively. The porosity of the prepared sample is found to be 49%.

3.2 Maximum Pore Radius

Bubble point method is used for the measurement of maximum pore radius of the bi-porous capillary wick. Here, the compressed air is allowed to pass through a wick immersed in a water tank. In the experiment, it is assumed that the wick is completely saturated with water, and all the pores are cylindrical in shape. When the compressed air is allowed to pass through the wick, the minimum pressure required to create the first bubble gives a measure of the largest pore radius. Figure 4 shows the experimental setup for calculating the maximum pore radius. Pressure at the inlet is recorded from the reading on the pressure gauge. The maximum capillary pore radius is then estimated using the Young–Laplace relation. The pore radius of the bi-porous wicks is calculated using Eq. (2) and is found to be around 85 nm. Here, σ presents the surface tension, and $r_{\mathrm{pore,\,max}}$ represents the maximum pore radius of the porous wick.

$$\Delta P = P_{\mathrm{guage}} - P_{\mathrm{atm}} = \frac{2\sigma}{r_{\mathrm{pore,\,max}}} \quad (2)$$

3.3 Permeability

The permeability of the capillary core is a measure of its ability to allow the fluid to flow under an applied pressure head given by Darcy's law. The permeability of the wick can be determined using Carman–Kozeny formula [16].

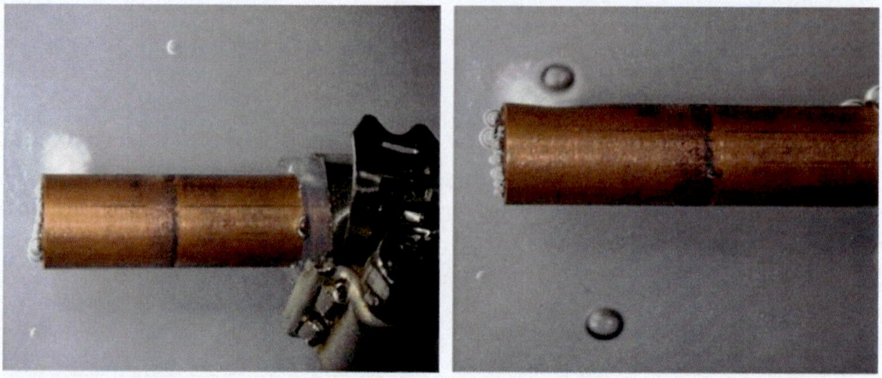

Fig. 4 Pore radius measurement using the bubble point method

$$k = \frac{d^2 \varepsilon^3}{180(1 - \varepsilon)^2} \quad (3)$$

The permeability of the fabricated bi-porous samples is found to be 0.0246 μm^2. In the above equation, k is the permeability, and 'd' is the pore diameter of the wick. Here, ε represents the porosity of the wick.

3.4 Surface Morphology

The surface morphology of the copper bi-porous wick is characterised by scanning electron microscopy (Gemini 300). Figure 5 shows SEM images of the sintered wick microstructure. SEM examination of the wick surface shows the existence of large pores, which leads to increased porosity and interconnects the fine pore network responsible for generating the required capillary pumping pressure.

3.5 Hardness Measurement

The hardness of the sample is a measure of its resistance to an indentation or localised deformation. It is desirable for the wick sample to have a moderate hardness which allows the ease of machining vapour grooves over its surface. The hardness of the prepared sample is measured using the Rockwell hardness test and is found to be 37.4 HRB (Rockwell Hardness Scale B).

Fig. 5 SEM images of sintered wick

4 Conclusion

In the current work, bi-porous copper wicks with distinguished shapes and sizes are successfully developed for loop heat pipes. The sintered copper wicks are characterised, and the properties like porosity, permeability, pore radius, surface morphology and hardness are calculated. An experimental setup is made for the measurement of maximum pore radius. The porosity of the prepared sample is found to be 49% using the density method. The pore radius of the bi-porous wicks is calculated using the Young–Laplace relation and is around 85 nm. The permeability of the fabricated bi-porous samples is found to be 0.0246 μm^2. SEM examination of the wick surface shows the existence of large pores, which leads to increased porosity and interconnects the fine pore network responsible for generating the required capillary pumping pressure. The measured hardness of the sintered sample is found to be 37.4 HRB.

Acknowledgements The characterisation experiments were conducted at the Central Instruments Facility of IIT Guwahati.

References

1. Ahmed S, Pandey M, Kawaji M (2022) Loop heat pipe design: an evaluation of recent research on the selection of evaporator, wick, and working fluid. J Therm Sci Eng Appl 14(7):070801
2. Ahmed S, Nashine C, Pandey M (2022) Thermal management at microscale level: detailed study on the development of a micro loop heat pipe. Micro Nano Eng:100150
3. Nashine C, Pandey M (2022) Theoretical modelling of miniature loop heat pipe. In: ASTFE digital library. Begel House Inc., 2022
4. Nashine C, Pandey M (2021) Thermo-hydraulic modelling of miniature loop heat pipes. In: Proceedings of the 26thnational and 4th international ISHMT-ASTFE heat and mass transfer conference, December 17–20, 2021, IIT Madras, Chennai, 600036, Tamil Nadu, India. Begel House Inc., 2021
5. Xu J, Zou Y, Yang D, Fan M (2013) Development of biporous Ti_3AlC_2 ceramic wicks for loop heat pipe. Mater Lett 91:121–124
6. Samanta SK, Das P, Lohar AK (2013) Study of physical characteristics of nickel wicks developed by metal injection moulding. Powder Metall 56(3):221–230
7. Singh R, Akbarzadeh A, Mochizuki M (2009) Effect of wick characteristics on the thermal performance of the miniature loop heat pipe. J Heat Transf 131(8)
8. Singh R, Akbarzadeh A, Mochizuki M (2009) Experimental determination of wick properties for loop heat pipe applications. J Porous Media 12(8)
9. Prasad A, Anand AR, Raghavendra Kumar D, Ramakrishnan V, Ambirajan A, Kumar D, Dutta P (2014) Measurement of thermal conductivity, pore size, permeability and coefficient of thermal expansion of porous nickel wick for lhps. Heat Pipe Sci Technol An Int J 5(1–4)
10. Wu S-C, Tzu-Wei G, Wang D, Chen Y-M (2015) Study of PTFE wick structure applied to loop heat pipe. Appl Therm Eng 81:51–57
11. Deng D, Liang D, Tang Y, Peng J, Han X, Pan M (2013) Evaluation of capillary performance of sintered porous wicks for loop heat pipe. Exp Thermal Fluid Sci 50:1–9
12. Siedel B (2014) Analysis of heat transfer and flow patterns in a loop heat pipe: modelling by analytical and numerical approaches and experimental observations. PhD diss., INSA de Lyon
13. Choi J, Yuan Y, Sano W, Borca-Tasciuc D-A (2014) Low temperature sintering of copper biporous wicks with improved maximum capillary pressure. Mater Lett 132:349–352
14. Choi J, Sano W, Zhang W, Yuan Y, Lee Y, Borca-Tasciuc D-A (2013) Experimental investigation on sintered porous wicks for miniature loop heat pipe applications. Exp Thermal Fluid Sci 51:271–278
15. Anvesh C, Srikanth T, Jasvanth VS (2021) Characterization of sintered porous wick for use in loop heat pipe. In: Proceedings of the 26th national and 4th international ISHMT-ASTFE heat and mass transfer conference, December 17–20, 2021, IIT Madras, Chennai-600036, Tamil Nadu, India. Begel House Inc., 2021
16. Dullien FAL (1991) Porous media: fluid transport and pore structure. Academic Press, New York

Laser-Induced Plasma-Assisted Ablation (LIPAA) of Transparent Materials

Upasana Sarma and Shrikrishna Nandkishor Joshi

1 Introduction

Microchannels are simple, recognisable patterns created to permit a regulated flow of fluids through them. The passages usually with a diameter below 1 mm are referred to as microchannels. Microchannels have widespread applications in the development of biomedical devices. The microchannel-based lab-on-a-chip [1–3] and microreactors [4] are some of the growing applications that can be used as customised diagnostic devices. Figure 1 shows some microchannel-based applications in real life.

Microchannels are fabricated on different materials like glass [5], metals [6] and polymers [7–9]. But then again, for all intents and purposes, microchannels demand its fabrication on materials with high transparency and high optical clarity. As such, machining of transparent materials with high precision is of great significance.

2 Transparent Materials in Scientific and Industrial Applications

Transparent materials like glass, polycarbonate (PC), polyimide (PI), polydimethylsiloxane (PDMS), polymethylmethacrylate (PMMA), etc. are the most trending transparent material. It is due to their eco-friendly nature and high transparency in the visible and near-IR spectral range [6]. Most transparent materials have exceptional resistance to chemical reactions and high thermal stability.

U. Sarma · S. N. Joshi (✉)
Department of Mechanical Engineering, Indian Institute of Technology Guwahati, Guwahati 781039, India
e-mail: snj@iitg.ac.in

S. N. Joshi et al. (eds.), *Low Cost Manufacturing Technologies*,
https://doi.org/10.1007/978-981-19-8452-5_11

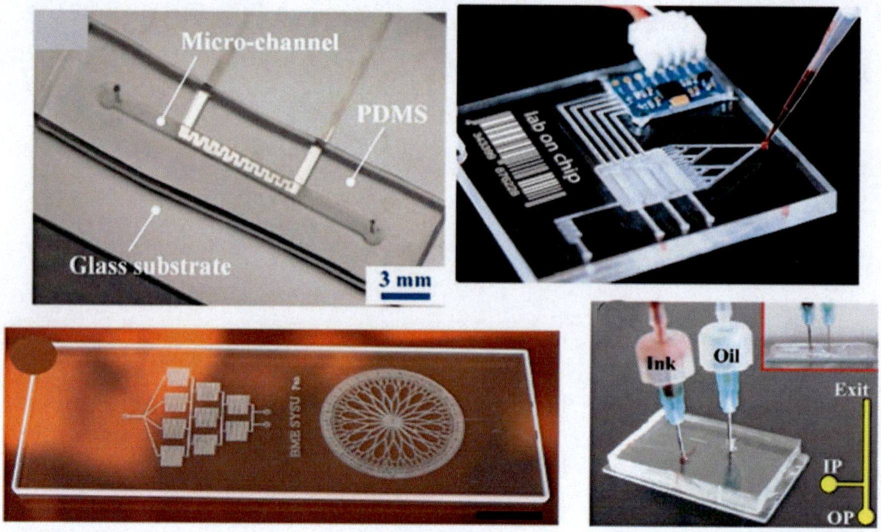

Fig. 1 Microchannel-based lab-on-a-chip set-ups [1, 2] (reproduced with permission from Elsevier)

Glass microfluidic devices provide exceptional chemical resistance, biocompatibility and optical properties, in addition to mechanical stability, which prevents swelling and deformation. After the initial focus on silicon, which is comparatively expensive and visually opaque to certain electromagnetic wavelengths, glass was chosen to build microfluidic chips, limiting its applications in optical detection [10]. Hnatovsky et al. [11] fabricated high-quality microchannels on fused silica and BK7 glass using femtosecond laser dielectric modification combined with selective chemical etching. Also, Rodriguez et al. [12] described two rapid and economic methods for fabricating microfluidic systems in glass. These glass microchannels were then applied in capillary electrophoresis where the robust surface characteristics of glass proves to be a convenient device. However, the expensive and time-consuming process for fabricating microchannels on glass and also the brittleness of glass restrict the development of glass-based microfluidic devices.

Because of the inherent impact strength, dimensional stability and optical clarity of PC, it is considered an ideal material for the medical device industry. Many researchers have reported microchannel fabrication in PC [13, 14]. Its ability to transmit light in a large range of wavelength makes it effective in visual monitoring of blood or other biological fluids. It is also used in developing Polycarbonate Urethane, (PCU) that acts as the bearing material in orthopaedic prosthesis, knee prosthesis [15], etc.

The unique properties of PDMS make it an important material for microchannels too. It is transparent and electrically and thermally insulating [8]. One can easily and affordably create microchips for biomedical purposes using PDMS-based microfluidic devices. Fujii [16] worked on PDMS-based microfluidic devices for applications, where microrcactors, microchips and hydrophobic vent valves are successfully integrated into the microchannels. Additionally, PDMS has qualities that make it a good substrate for miniature biological devices by fabricating PDMS microstructures [17].

PMMA, also known as acrylic or acrylic glass, is a transparent and rigid thermoplastic material widely used as a shatterproof replacement for glass. It is used in biomaterial applications such as bone cement, lenses, bone substitutes and drug delivery systems. Smooth walled cylindrical microchannels with diameters in the range of 8–20 μm were fabricated on PMMA substrate by Day and Gu [18]. Dudala et al., [19] and Prakash and Kumar [20] also fabricated microchannels on PMMA substrate using the CO_2 laser for microfluidic applications. Further, PMMA microfluidic device with filtration features was used to separate red blood cells from blood [21]. Microchannels in PMMA substrates bonded to porous PETE track-etched membranes are also utilised for the cytotoxicity testing of anticancer drugs in a microfluidic device [22].

3 Microchannel Fabrication Technologies

Microchannels can be generated on substrates by using conventional as well as unconventional manufacturing processes, viz. microwire moulding, imprinting, lithography, chemical etching, plasma etching and laser-based machining. These techniques are capable of machining microchannels on a variety of materials; however, some of these have certain limitations. Due to the degassed polymer spilling out of the mould, more waste is generated during microwire moulding. Additionally, the technique is expensive and time-consuming because to the complexity of the mould and its operation. Whereas, in lithography, the substrate often absorbs the developing solution during the production of microchannels. Its need for a chamber that blocks UV light also makes it an expensive operation. In the imprinting approach, the delayed mass movement and ongoing material entrapment in the metal blocks result in a non-uniform distribution of material. However, if laser parameters like wavelength, laser intensity, pulse duration, etc. are properly tuned, laser direct machining has been proven to be a promising tool. On opaque materials, laser processing has demonstrated its capacity to create neat and uniform channels [23].

4 Laser-Based Micromachining

In laser-based micromachining, the material is removed from the substrate using thermal energy that is derived from the light energy of a highly coherent stream of photons. The substrate is exposed to a laser beam, which is mostly absorbed and some of which is reflected. The energy that is absorbed is instantly transformed into heat energy, raising the temperature in the area that has been exposed to radiation. The substance melts and vaporises when the laser energy is high enough. The vapour that follows from the vaporisation creates a recoil pressure at the melt pool's surface and causes the melt to be ejected. As such, this process works effectively with materials that are capable of absorbing laser energy.

4.1 Effect of Workpiece Material Properties

During laser-based machining, the material ablation is a thermal process. The temperature distribution is mainly responsible for the desired process outcome. The thermal and optical properties of the workpiece materials have a major effect on the laser ablation process. The important materials properties are discussed below.by lip

Optical property

An optical property of a substance called absorptivity specifies how much light may be absorbed in a material, proportion to the amount of light that is incident on it. The wavelength and direction of the incident light, the kind of material, its chemical content and physical structure, as well as the condition of the material and its surface, all affect the absorptivity value. Absorptivity (A) can be calculated as:

$$A = 1 - R - T \tag{1}$$

where R is the reflectivity and T is the transmissivity of the material.

A workpiece surface does not totally absorb the laser beam. A portion of the beam is absorbed and the remaining portion is reflected back into the environment. Increase in absorptivity increases the laser energy input into the workpiece surface. In a time period of around 10^{-13} s, the absorbed energy stimulates the free electrons, which instantly transforms into heat [24]. The heat is subsequently disintegrated through a variety of heat transmission mechanisms, including conduction, convection and radiation, but conduction has the most significant impact [25]. The heat conducted into the metal surface raises the surface temperature, assisting the laser ablation process. Lawrence [26] discovered that a high-power diode laser is more reliable than a CO_2 laser because the diode laser's wavelength is small, which increases absorption because most materials are greatly absorbing at low wavelength. Similarly, Li et al. [21] examined the effect of the absorptivity of metal on femtosecond pulsed laser ablation. It was observed that the absorptivity of the material increases with the temperature evolution with time and as such, an increase in the ablation is achieved.

Transparent materials, on the other hand, are highly transmissible to laser beams over a wide wavelength range. The laser beam passes through the transparent material without disruptions, resulting in no thermal effect on the material. Researchers from all over the world have used CO_2, excimer and Nd: YAG lasers to fabricate microchannels in transparent materials like glass [27] and polymers [18]. It was also discovered that CO_2 and excimer lasers could effectively machine the transparent polymer material because most polymers have significant absorptivity in the far IR spectrum and low UV wavelength [20].

Thermal property

The temperature distribution in a workpiece is controlled by its thermal properties. Thermal conductivity (k) and heat capacity (c) are the important thermal properties. The thermal conductivity of metals is such a thermal property that signifies its ability to conduct heat into the material. In metals, the value of thermal conductivity varies in a wide range, from a value of approximately 8 W/mK to a value of about 400 W/mK. Thermal conductivity affects the thermal gradient and peak temperature because it determines heat transfer into the surrounding material [28]. Peak temperature and temperature gradient both decrease as thermal conductivity increases. It is because of the rapid heat dissipation in high conductivity materials [29]. Benton et al. [30] studied the effect of thermal conductivity on laser micromachining of microchannels and observed that the melted and vaporised volume in the metal target increases with the increase in the thermal conductivity value of the material.

Specific heat capacity on the other hand signifies the amount of heat required to be supplied per unit mass to raise the temperature by unit degree. In metals, its value range from a lowest value of about 120 J/kgK to a highest value of 3500 J/kgK. Similar to thermal conductivity, specific heat also governs the peak temperature obtained during the laser irradiation. In reality, the specific heat of a material governs the development of the peak temperature when various materials absorb the same amount of heat. The smaller the specific heat is, the higher the peak temperature is obtained [31]. Benton et al. [30] also investigated the effect of specific heat capacity on laser micromachining of microchannels and reported that, the higher the specific heat of a metal, the lower the depth of cut for a given laser power is obtained.

From the aforementioned studies, it can be observed that depending on the various factors such as wavelength of the incident laser beam, surface roughness of the material and the temperature evolution on the surface, the metals get highly absorbing thereby causing thermal ablation on its surface. Transparent materials exhibit significant absorptivity at the far IR spectrum and low UV wavelength. No laser beam can be absorbed in the visible and near-IR spectral range making the laser machining of transparent materials in such spectral range a challenging task. It may be worthy studying the laser machining process of transparent materials in the visible and near-IR spectral range.

4.2 *Motivation for LIPAA Process*

The laser must be strongly absorbed by the sample when fabricating microchannels with lasers. But nevertheless, most microchannels necessitate the use of transparent materials. Transparent materials with high transparency do not absorb laser energy, making laser machining difficult. Nonetheless, ultrashort lasers can process transparent materials because they generate a highly concentrated light field that is absorbed into the transparent material via nonlinear absorption and is powerful enough to ionise any atom. Short wavelength lasers, like ultrashort lasers, can also process transparent materials. Thus, it is rather challenging to perform laser processing on transparent materials using long-pulsed, long-wavelength lasers. However, it should be mentioned that at such condition, the LIPAA technique is capable of processing transparent materials.

5 Theoretical Background of LIPAA Process

LIPAA involves placing an extra absorbent layer below the transparent material. There are three categories for the LIPAA process depending on the absorbing layer's material (i) Laser-Induced Backside Wet Etching (LIBWE), (ii) Laser-Induced Backside Dry Etching (LIBDE) and (iii) Laser Etching at a Surfaced-Adsorbed Layer (LESAL). Under the transparent material in LIBWE, a highly absorbing solvent is being employed. It is distinguished by a low rate of etching, a smooth surface and an incubation effect [32]. Whereas, by continually adsorbing hydrocarbon by-products from a gaseous organic medium, LESAL creates a thin absorbent layer on the backside of the transparent material [33]. Likewise, a thin metal is positioned below the transparent material in LIBDE. It is distinguished by a fast etching rate and the absence of incubation effects [34]. Figure 2 shows the principal experimental set-up for the three categories of LIPAA.

LIBDE on the other hand is often denoted by the general term LIPAA. In the LIPAA process, plasma is generated on the rear side of the transparent material, assisting the ablation of the material on its rear side. In this process, the transparent material is placed above a metal target sheet with no pressure being exerted on the transparent material. The schematic representation of the LIPAA experimental set-up and its mechanism is shown in Fig. 3. A detailed flow chart describing the mechanism of LIPAA is also being shown in Fig. 4.

The transparent material is translucent to a wide range of laser wavelength. Thus, the incoming laser beam passes through the transparent material and is focused on the upper surface of the metal target sheet. A fraction of the laser energy is absorbed by the absorbing metal target sheet. The absorbed energy results in the excitation of the free electrons, which instantaneously gets converted into heat energy [24]. This heat energy then gets dissipated through various modes of heat transfer, viz. conduction, convection and radiation. Conduction plays a vital role in conducting the energy into the metal surface, thereby increasing the temperature generation on the metal target. Once the intensity of laser is high enough ($> 10^5$–10^8 W/cm^2), the temperature generated on the metal target sheet surpasses its melting and boiling point temperature. The vapour thus formed from the metal surface contains a cluster of molecules. Subsequently, the interactions of the laser beam with the vapour lead to the ionisation of the vapour molecules resulting in the formation of opaque and dense plasma. The plasma generated again interacts with the laser beam, resulting in Inverse Bremsstrahlung absorption and leads to an expansion of the plasma [1]. The plasma now has a significant thermal effect on both the transparent and the metal target sheet, thereby ablating both the materials.

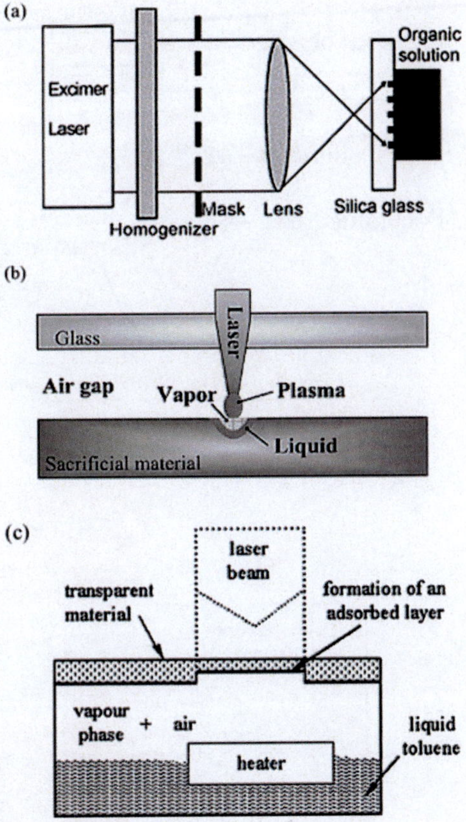

Fig. 2 Principal experimental set-up for **a** LIBWE, **b** LIBDE and **c** LESAL [1, 35, 36] (reproduced with permission from Elsevier)

6 Fabrication of Microchannels on PC By LIPAA

Using the LIPAA process, microchannels were successfully fabricated in our laboratory on transparent polycarbonate (PC) sheet. A conventional solid-state Nd: YAG pulsed laser of 1064 nm wavelength and pulse width in the millisecond range has been utilised for microchannel fabrication. The experimental set-up of the LIPAA process is presented in Fig. 5. The transparent material is a PC sheet having a thickness of 1 mm, while the metal target taken is a sheet of copper of 1 mm thickness.

The microchannel fabrication on PC using LIPAA is carried out by varying three laser parameters, viz. pulse repetition rate (PRR), pulse width (PW) and pulse power density (PPD). The microchannel dimensions are measured and analysed form FESEM images. The FESEM images of a microchannel created by LIPAA are shown in Fig. 6 under the process conditions of 60 Hz PRR, 4 ms PW and a 3.055 MW/cm^2 PPD.

The investigation reveals that when the laser parameters increases, the channel dimensions also rise. It is because more stimulated emission occurs with a higher pulse power density. As a result, increased photon exposure on the metal sheet causes

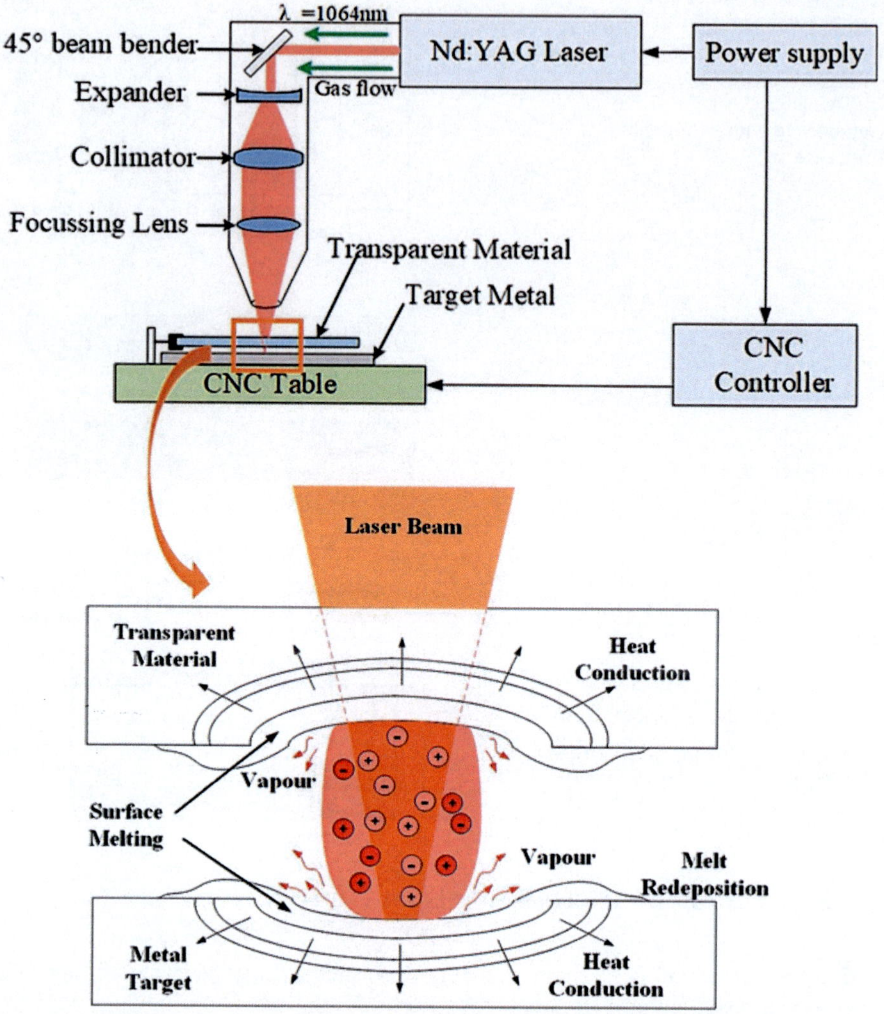

Fig. 3 Schematic representation of the experimental set-up and mechanism of LIPAA

a rise in the electron and ion number density. It is known that, generation of plasma is governed by the electron and ion density, increasing which will result in an enriched plasma generation [37]. Similar to this, the quantity of laser pulses engaging with the metal sheet grows as the pulse repetition rate does. In turn, the molecules in the vapour become more vaporised and ionised. Additionally, an increase in plasma synthesis is correlated with an increase in pulse duration. As a result, plasma formation increases along with the laser parameters. The polycarbonate sheet is thermally affected by this enhanced plasma, increasing the channel dimension.

A layer of carbon particles with some minute traces of copper and oxygen particles also gets deposited on the edge and bed of the microchannel. However, the deposited

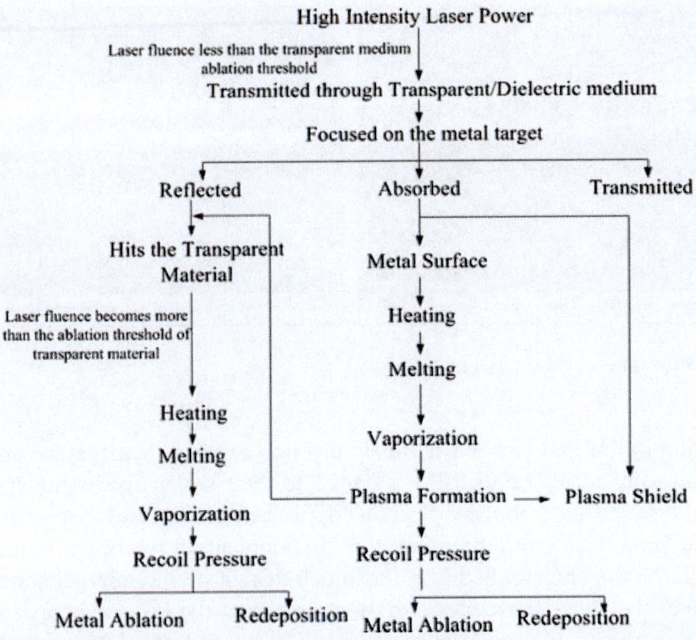

Fig. 4 Mechanism of LIPAA

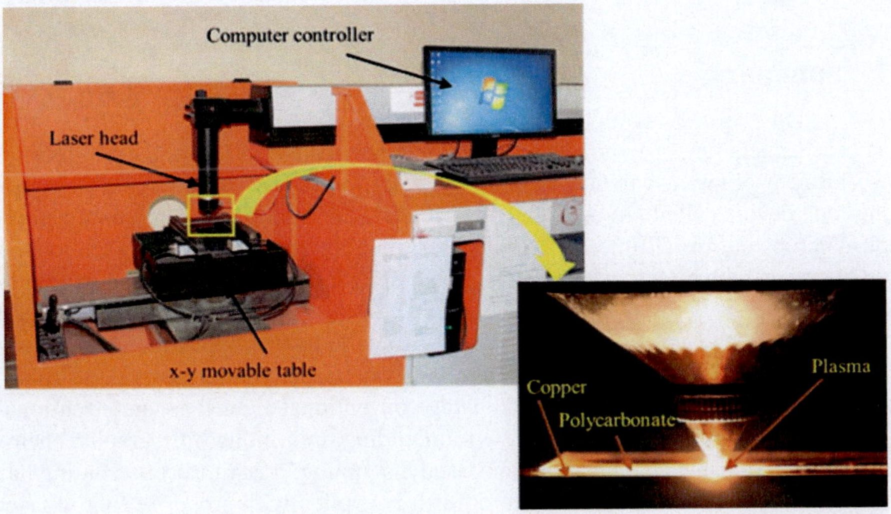

Fig. 5 Nd: YAG laser machine set-up for LIPAA process

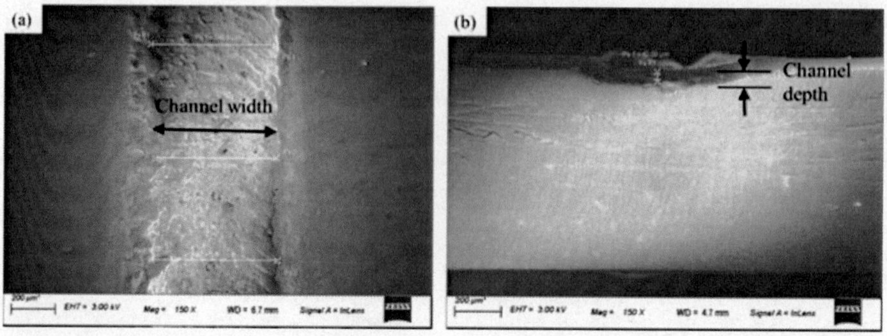

Fig. 6 FESEM image of **a** C_{width} and **b** C_{depth}

carbon, aluminium and oxygen particles are not permanent in nature and can be cleaned ultrasonically. The FESEM images and the EDX analysis shown in Fig. 7. Prove the presence of carbon, oxygen and copper particles on and around the channel bed. It can be noticed from the figure that, the deposition of copper particles from metal target on the channel bed is not as much as that on the microchannel edge. It is because of the shock wave at the irradiated zone of the PC and copper sheet that leads to an explosive melt expansion resulting in deposition of the melted material on the channel edge.

7 Summary

Microchannels have become an indispensible part of micro technology in today's world. They allow easy transportability of devices and incorporate several processes into one device called lab-on-a-chip. Microchannels can be fabricated on a variety of materials but for all intents and purposes, microchannels demand its fabrication on materials with high transparency and high optical clarity. Most transparent materials have a very good resistance to chemical reactions, high thermal stability and high hardness and as such, find wide scientific and industrial applications. Materials like glass, PC, PI, PDMS, PMMA, etc. are the most trending transparent material. Microchannels can be generated using conventional as well as unconventional manufacturing processes, viz. microwire moulding, imprinting, lithography, chemical etching, plasma etching and laser-based machining. Laser direct machining has however been found as a promising tool if it is applied with proper setting of laser parameters, viz. short wavelength and ultrashort pulse. The laser must be highly absorbed by the material. But high transparency of the transparent material does not allow laser energy to be absorbed at longer wavelength lasers and longer pulses. Hence, it is difficult to machine transparent material by a laser. LIPAA however has the potential to process transparent materials. As such, a brief discussion on microchannel fabrication on PC using LIPAA has been presented in the current

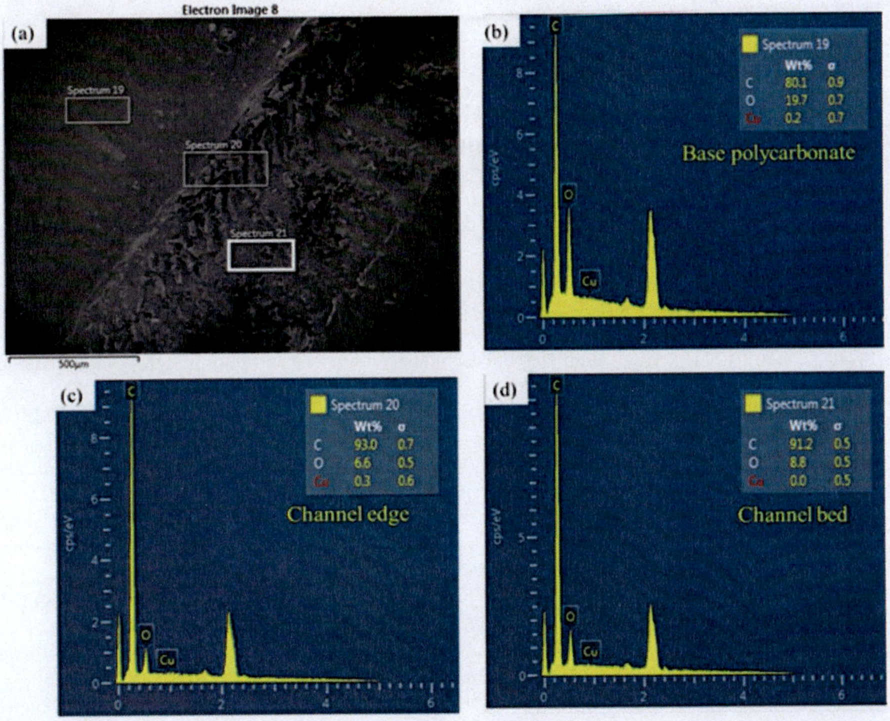

Fig. 7 EDX analysis showing the elemental composition at **a** base polycarbonate, **b** channel edge and **c** channel bed

chapter. Further, minute traces of copper and oxygen particles deposited on the edge and bed of the microchannel have also been analysed. The deposits are however not permanent in nature and can be cleaned ultrasonically.

References

1. Pan C, Chen K, Liu B, Ren L, Wang J, Hu Q, Liang L, Zhou J, Jiang L (2017) Fabrication of micro-texture channel on glass by laser-induced plasma-assisted ablation and chemical corrosion for microfluidic devices. J Mater Process Technol 240:314–323
2. Xu S, Liu B, Pan C, Ren L, Tang B, Hu Q, Jiang L (2017) Ultrafast fabrication of micro-channels and graphite patterns on glass by nanosecond laser-induced plasma-assisted ablation (LIPAA) for electrofluidic devices. J Mater Process Technol 247:204–213

3. Kim HG, Park MS (2017) Circuit patterning using laser on transparent material. Surf Coat Technol 315:377–384
4. Suryawanshi PL, Gumfekar SP, Bhanvase BA, Sonawane SH, Pim-plapure MS (2018) A review on microreactors: reactor fabrication, design, and cut-ting-edge applications. Chem Eng Sci 189:431–448
5. Nieto D, Delgado T, Flores-Arias MT (2014) Fabrication of microchannels on soda-lime glass substrates with a Nd: YVO4 laser. Opt Lasers Eng 63:11–18. https://doi.org/10.1016/j.optlaseng.2014.06.005
6. Singh SS, Baruah PK, Khare A, Joshi SN (2018) Incubation studies and the threshold for surface damage and cavity formation in the processing of polycarbonate by Nd: YAG laser. Opt Laser Technol 108:592–601
7. Kim KR, Kim HJ, Choi HI, Shin KS, Cho SH, Choi BD (2015) Ultrafast laser microfabrication of a trapping device for colorectal cancer cells. Microelectron Eng 140:1–5
8. Li G, Xu S (2015) Small diameter microchannel of PDMS and complex three-dimensional microchannel network. Mater Des 81:82–86
9. Prakash S, Kumar S (2017) Fabrication of rectangular cross-sectional microchannels on PMMA with a CO_2 laser and underwater fabricated copper mask. Opt Laser Technol 94:180–192
10. Hwang J, Cho YH, Park MS, Kim BH (2019) Microchannel fabrication on glass materials for microfluidic devices. Int J Precis Eng Manuf 20(3):479–495
11. Hnatovsky C, Taylor RS, Simova E, Rajeev PP, Rayner DM, Bhardwaj VR, Corkum PB (2006) Fabrication of microchannels in glass using focused femtosecond laser radiation and selective chemical etching. Appl Phys A 84(1):47–61
12. Rodriguez I, Spicar-Mihalic P, Kuyper CL, Fiorini GS, Chiu DT (2003) Rapid prototyping of glass microchannels. Anal Chim Acta 496(1–2):205–215
13. Chen X, Hu Z (2017) An effective method for fabricating microchannels on the polycarbonate (PC) substrate with CO laser. Int J Adv Manuf Technol 92
14. Qi H, Chen T, Yao L, Zuo T (2009) Micromachining of microchannel on the polycarbonate substrate with CO_2 laser direct-writing ablation. Opt Lasers Eng 47(5):594–598
15. Kanca Y, Milner P, Dini D, Amis AA (2018) Tribological evaluation of biomedical polycarbonate urethanes against articular cartilage. J Mech Behav Biomed Mater 82:394–402
16. Fujii T (2002) PDMS-based microfluidic devices for biomedical applications. Microelectron Eng 61:907–914
17. Sia SK, Whitesides GM (2003) Microfluidic devices fabricated in poly (dimethylsiloxane) for biological studies. Electrophoresis 24(21):3563–3576
18. Day D, Gu M (2005) Microchannel fabrication in PMMA based on localized heating by nanojoule high repetition rate femtosecond pulses. Opt Express 13(16):5939–5946
19. Dudala S, Rao LT, Dubey SK, Javed A, Goel S (2020) Experimental characterization to fabricate CO_2 laser ablated PMMA microchannel with homogeneous surface. Mater Today: Proc 28:804–807
20. Prakash S, Kumar S (2015) Fabrication of microchannels on transparent PMMA using CO_2 Laser (10.6 μm) for microfluidic applications: an experimental investigation. Int J Precis Eng Manuf 16(2):361–366
21. Li JM, Liu C, Dai XD, Chen HH, Liang Y, Sun HL, Tian H, Ding XP (2008) PMMA microfluidic devices with three-dimensional features for blood cell filtration. J Micromech Microeng 18(9):095021
22. Nguyen T, Jung SH, Lee MS, Park TE, Ahn SK, Kang JH (2019) Robust chemical bonding of PMMA microfluidic devices to porous PETE membranes for reliable cytotoxicity testing of drugs. Lab Chip 19(21):3706–3713
23. Majumdar JD, Manna I (2003) Laser processing of materials. Sadhana 28(3):495–562
24. Dahotre NB, Harimkar S (2008) Laser fabrication and machining of materials. Springer Science & Business Media
25. Mishra I, Lattanzi AM, LaMarche CQ, Morris AB, Hrenya CM (2019) Experimental validation of indirect conduction theory and effect of particle roughness on wall-to-particle heat transfer. AIChE J 65(10):e16703

26. Lawrence J (2002) A comparative investigation of the efficacy of CO_2 and high-power diode lasers for the forming of EN3 mild steel sheets. Proc Inst Mech Eng Part B: J Eng Manuf 216(11):1481–1491
27. Chang TL, Chen ZC, Lee YW, Li YH, Wang CP (2016) Ultrafast laser ablation of soda-lime glass for fabricating microfluidic pillar array channels. Microelectron Eng 158:95–101
28. Bejan A, Kraus AD (eds) (2003) Heat transfer handbook, vol 1. John Wiley & Sons
29. Hu Z, Kovacevic R, Labudovic M (2002) Experimental and numerical modeling of buckling instability of laser sheet forming. Int J Mach Tools Manuf 42(13):1427–1439
30. Benton M, Hossan MR, Konari PR, Gamagedara S (2019) Effect of process parameters and material properties on laser micromachining of microchannels. Micromachines 10(2):123
31. Guan Y, Sun S, Zhao G, Luan Y (2005) Influence of material properties on the laser-forming process of sheet metals. J Mater Process Technol 167(1):124–131
32. Zimmer K, Böhme R (2008) Laser-induced backside wet etching of transparent materials with organic and metallic absorbers. Laser Chem
33. Ehrhardt M, Lorenz P, Zimmer K (2012) Surface modification by laser etching using a surface-adsorbed layer. Thin Solid Films 520(9):3629–3633
34. Hopp B, Vass C, Smausz T (2007) Laser induced backside dry etching of transparent materials. Appl Surf Sci 253(19):7922–7925
35. Ding X, Kawaguchi Y, Sato T, Narazaki A, Kurosaki R, Niino H (2004) Micron-and submicron-sized surface patterning of silica glass by LIBWE method. J Photochem Photobiol, A 166(1–3):129–133
36. Böhme R, Zimmer K (2004) Low roughness laser etching of fused silica using an adsorbed layer. Appl Surf Sci 239(1):109–116
37. Sarma U, Joshi SN (2020) Machining of micro-channels on polycarbonate by using laser-induced plasma assisted ablation (LIPAA). Opt Laser Technol 128:106257

CO_2 Laser Cutting of White Pat Silk—A Preliminary Work

Evenmore Mylliem and Shrikrishna Nandkishor Joshi

1 Introduction

Laser has been employed tremendously in the garment and apparel industries in processing different kinds of fabric materials. The main reason for the utilization of lasers in the garment and apparel industries is their flexibility and their ability to reduce the cost of production in the long run. Lasers have many applications in the garment and apparel industries ranging from cutting, engraving, marking, patterning, fading and many other applications. The ease of integration of the laser with the computer system has led to its increased utilization in recent years.

1.1 Laser Technology

Laser is an acronym for "Light Amplification by Stimulated Emission of Radiation." The word "laser" refers to both a device and a technology. The invention of the first working laser in the year 1960 by Theodore Maiman using ruby as an active medium has opened the way for the invention of lasers of different wavelengths using the different types of materials as the active media. Laser is electromagnetic radiation produced by the atoms by the change in their energy states whenever the materials are being excited. The excited atoms emit laser in the form of light, and this laser light is further amplified in a suitable lasing medium equipped with a specific arrangement of mirrors. The laser was finally produced from the arrangement as a stream of

E. Mylliem · S. N. Joshi (✉)
Department of Mechanical Engineering, Indian Institute of Technology Guwahati, Guwahati, Assam 781039, India
e-mail: snj@iitg.ac.in

E. Mylliem
e-mail: emylliem@iitg.ac.in

S. N. Joshi et al. (eds.), *Low Cost Manufacturing Technologies*,
https://doi.org/10.1007/978-981-19-8452-5_12

light having a color determined by its wavelength. The laser light produced has four main properties: intensity, monochromaticity, coherency and collimation which differentiates it from natural light. The basic components of any laser system include the active or lasing medium, energy or pumping source and the optical resonator. Lasers are classified into gas lasers, solid-state lasers, dye lasers and semiconductor lasers based on the lasing medium.

The different process parameters which play an important role on laser-material interaction include wavelength, laser power, scan speed, laser pulse width, laser pulse frequency, laser optics and material properties. These parameters need to be wisely controlled or selected to achieve better output results in the laser processing of any materials. Laser has gained popularity in the garment and apparel industries because it is a non-contact and force-free process, its power and speed can be easily controlled, and it can be easily integrated with the computer systems therefore complex shapes and patterns can be cut from the fabric materials.

1.2 Fabric

Fabric refers to the flexible material or structure, formed by the interlacing or interlocking of yarns or threads. The yarns and threads are produced by the spinning of raw fibers (obtained from natural or synthetic products) into long and twisted lengths. The process of interlocking and interlacing yarns or threads is usually known as weaving, knitting, felting and bonding. The fibers are classified into natural and man-made fibers depending on the source from which they are extracted. The fabric can be made up of purely natural fibers or purely synthetic fibers or a mixture of both two.

The process of rearing silkworms for the production of raw silk fibers is known as sericulture. Pat silk is a natural fabric obtained from the cocoons of the mulberry silkworms. Assam a state in North-East India is famous for producing different types of silk like Pat silk, Muga silk, Tasar silk and Eri silk. Sualkuchi in Assam is well-known for producing different traditional textiles made from different silk fibers, and it is popularly known as the "Manchester of Assam." Pat silk apart from being known as the expensive along with Muga silk and also has a major share in the silk industry in Assam. Pat silk is well-known for its durability and glossy texture and has a natural white tint color. The pat silk, like other silks produced in Assam, has been used in making products like mekhelas, chadars and many other useful textiles.

2 Literature Review

Laser cutting is one of the many processes that can be carried out using laser technology. Laser cutting has been used to cut metals and many non-metals accurately into different types of regular and complex shapes. In the garment industries, the use of lasers for cutting is very important as many shapes and patterns from the fabric rolls can be cut to required sizes, thereby reducing the waste of fabric materials. From the literature, we have seen that the continuous wave CO_2 laser was used widely for cutting and processing different fabric materials. CO_2 laser has been used as a high-speed cutting tool in the precise cutting of polymer-based fabrics in a defined shape and the fibers melt and stick with each other to prevent fraying [1, 2]. CO_2 laser cutting of single-layer and multiple-layer fabrics obtained from both natural and synthetic products has shown the need for proper control of the laser process parameters, i.e., laser power and speed, so that good cutting results can be obtained. Laser cutting of multiple layers of fabrics made up of natural fibers results in excessive burning, but in the case of synthetic fabrics, the edges of the layers of fabrics stick to one another due melting of the fibers at the cutting edges [3]. The laser process parameters, namely laser power and speed, are the main parameters that need to be properly selected to achieve an accurate cutting edge with less damage [3–5]. Proper combinations of laser parameters are needed to ensure a proper cut through the felt fabric [6].

There are many other laser processing techniques apart from cutting that are employed for the processing of different garments. The famous among these laser techniques is laser engraving. The use of a laser to modify and create patterns on the fabric surface without a through cut of the surface is called laser engraving. For laser engraving, pulsed CO_2 laser with pixel time and resolution being the process parameters mostly controlled. Laser engraving has been carried out by many researchers and reported that when proper combination of parameters is selected during laser processing desired results can be achieved on the fabric without damage to the fibers. Varying the irradiation intensity (a parameter which is directly proportional to maximal power, duty cycle and pixel time) during laser engraving of dyed and undyed 100% cotton fabrics gave rise to change of color on the surface of the fabrics as reported by Štěpánková et al. [7] and adjusting the proper range of values for irradiation intensity can prevent unwanted damage to the laser processed fabrics. Using laser engraving, a faded and worn out look on the denim fabrics was created which suggested that laser engraving is the best alternative to different traditional denim finishing techniques as there is no use of water and chemicals during the process [8, 9]. Laser engraving of fabrics not only creates color changes, but it is also associated with reduction in the tensile strength and breaking strength of the fabrics and improve wettability when processed with proper combination of process parameters [10–12]. The use of graphic design method and resist design method in combination with laser engraving has imparted many complex artistic designs on the processed fabrics and the color appearances of the design on the fabrics can be easily managed by controlling the laser treatment parameters [13–15]. The use of laser engraving in creating artistic designs is more environmentally friendly as no water and chemicals are utilized. Metallic appearance with patterns was imparted on the denim by laser engraving of the aluminum foil laminated on the fabric [16]. Laser treatment of fabrics alter the surface properties fabrics, thereby exposing unsaturated bonds and free radicals on the surface which further improve the dye absorption of the treated portion of the fabric as compared to the untreated portion resulting in different color intensity after dyeing [17–20]. Laser treatment of fabric before dyeing can improve the color fastness and light fastness of the fabric after dyeing [21]. The use of laser engraving technique on cotton fabric has led to the improvement of the fibers to absorb more of silver nanoparticles as compared to the untreated, thereby improving the antibacterial properties of the fabric even after washing [22]. To control felting shrinkage of wool, laser treatment with appropriate parameters was conducted and it is reported that laser treatment can be an alternative to traditional chlorination method to control the felting shrinkage in wool [23]. Laser treatment of fabric also modifies the fibers on the surface, thereby leading to reduction in the pilling of fibers in the surface of the fabrics [24].

From the literature, it can be seen that there exists an enormous number of papers in the area of laser engraving but a limited number of papers reporting laser cutting of different fabrics. Pat silk, obtained from mulberry silkworms which are famous in the state of Assam situated in the North-Eastern region of India has not been reported to be processed by laser. Therefore, this work was carried out to establish a laser cutting process of pat silk with a 2.5 kW CO_2 laser. The result of this work is satisfying and further modifications can be made to improve the processing of pat silk.

3 Methodology

The pat silk fabric is a flexible material, and to process it by the selected 2.5 kW CO_2, it is necessary to fix it onto a rigid material so that processing can be done accurately. To fulfill this condition, the white pat silk fabric was taped onto the supporting base material made up of a rubber mat (5 mm thickness) and a cartoon box sheet (2 mm thickness). The rubber mat was taped on the top of the cartoon box sheet, and the cartoon box sheet provides the necessary rigidity.

To carry out the laser cutting using a 2.5 kW CO_2 laser, a circle of 5 mm diameter was designed using AutoCAD and the design was transferred to the CADMAN-PL for the generation of the toolpath. After the generation of the toolpath, the CNC program was generated and then loaded into the CO_2 laser machine memory so that the movement of the laser cutting head can be controlled while carrying laser cutting of the pat silk. The CO_2 laser was then prepared for the process and necessary conditions have been set so that accurate cutting can be obtained.

The following conditions have been considered for carrying out the laser cutting (Table 1).

Table 1 Conditions for CO_2 laser cutting of white pat silk

Laser power	100 W (minimum)
Scan speed	7000 mm/min (maximum)
Stand-off-distance	1 mm
Assisting gas	Nitrogen, N_2
Assisting gas pressure	0 to 5 bars, with an increment of 1 bar

4 Results and Discussion

The preliminary experimentation is conducted in 2.5 kW CO_2 Laser (Make: LVD). The pat silk fabric is mounted using mounting tape to fix it onto the supporting substrate consisting of rubber mat and cartoon box sheet as shown in Fig. 1.

The experimentations were carried out by varying the pressure of the assisting gas from 0 to 5 bars with an increment of 1 bar and keeping other parameters like laser power and scan speed constant at 100 W and 7000 mm/min, respectively. As observed from Fig. 2a, b, it can be seen that some of the circles are laser cut fully and some have fibers still attached to parent silk fabric. This observation suggests that due to the presence to the presence of assist gas during the laser cutting there might be a possibility that the fabric is blown and owing to this blowing the part of the fabric along the laser cutting path may be out of focus leading to some of the fibers being attached to the parent fabric even after laser has passed through the cutting path.

It can be also observed from Fig. 2a, b that the circle being laser cut has more overcut when the assist gas pressure was set to 0, 1, 2, 3 bars. It is also observed that the blackening of the laser cut edge is more when the assist gas pressure was set to 0, 1, 2, 3 bars. The reason for large overcut and blackening of the laser cut edge may be due to the oxidation from the surrounding during the laser cutting. The overcut is observed to be largely reduced when the assist gas pressure was set to 4 and 5 bars and with same values of assist gas pressure the blackening of the laser cut edge is almost not noticeable. As the assist gas pressure is increased, the laser beam is constricted from the surrounding oxygen, thereby reducing the chances of oxidation from surrounding.

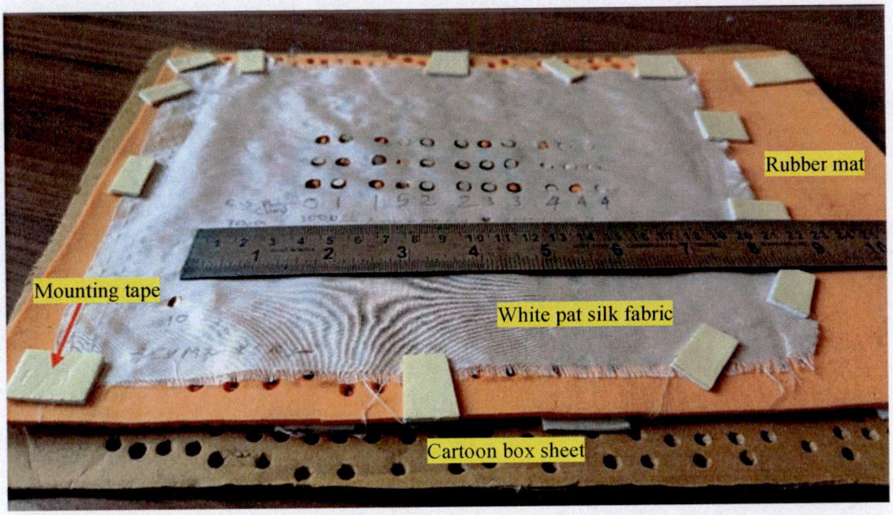

Fig. 1 Laser processed white pat silk fabric mounted on the supporting substrate of rubber mat and cartoon box sheet

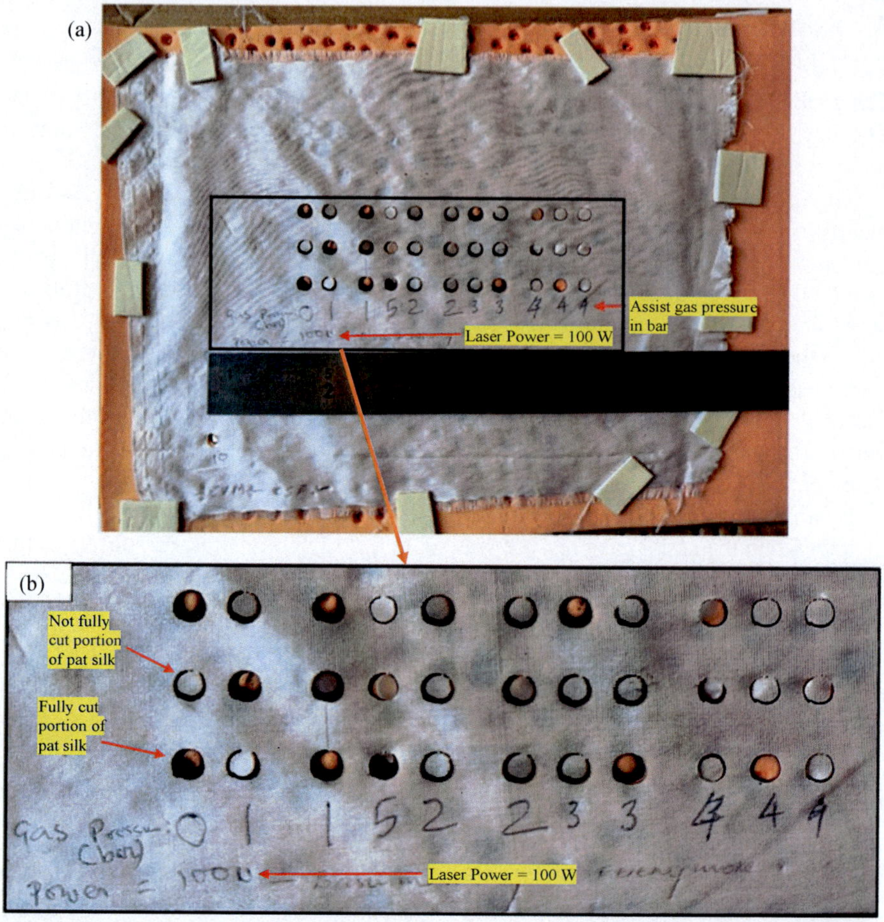

Fig. 2 **a** Laser cutting of pat silk at laser power = 100 W, scan speed = 7000 mm/min and assist gas pressure from 0 to 5 bars with 1 bar increment. **b** Enlarged view of the laser processed portion

From the preliminary experimentation, it is also observed that using the laser power of 100 W the laser can even penetrate through the rubber mat up to the cartoon box sheet. This is due to low melting temperature rubber and low burning temperature of the cartoon box material. This observation suggests the use of substrate material having higher melting temperature-which can resists penetration of the laser through it.

In this preliminary experimentation, the main process parameters are laser power, scan speed and assist gas pressure. As discussed above when laser power and scan speed are kept constant, varying the assist gas pressure has given rise to different effects of the laser on the laser cut edge. The fabric being a thin and flexible material needs to be properly held in its place so that the laser processed results can be improved further.

5 Conclusion

The following conclusions can be drawn from the results obtained after processing white pat silk with a 2.5 kW CO_2 laser:

- When pat silk was laser cut, the edge of the laser processed portion was observed to have a less blackening effect due to burning when the assist gas was set to the value of 4 bars and 5 bars with other parameters set to selected constant values as mentioned in Table 1.
- The work suggests that a laser machine with lower minimum power can be a good choice for obtaining good results during the experimentation on the selected fabric.
- Laser cutting of the fabric can prevent the fraying of fibers that form the fabric.

References

1. Yilbas BS (1986) Cloth cutting by carbon-dioxide laser, no September, pp 143–145
2. Jackson M, Preston M, Yang S, Tao L (1994) Laser cutting parameters for high-speed cutting of net fabrics. Laser Mater Process Mach 2246(Nov 1994):56–66. https://doi.org/10.1117/12.193123
3. Yusoff N, Osman NAA, Othman KS, Zin HM (2010) A study on laser cutting of textiles. In: 29th International congress on applications of lasers and electro-optics, ICALEO 2010—Congress Proceedings, 2010, vol 103, no January, pp 1559–1566. https://doi.org/10.2351/1.5062018
4. Dolchinkov N, Shterev Y, St L, Boganova D, Peneva M, Linkov L, Nedialkov D (2019) Exploring the possibility of laser cutting with CO_2 laser on felt in the range from 1W to 26W power. Industry 4.0 4(1):29–31
5. Saaidin H, Tajuddin RM (2018) Denim casual wear with laser cut Pucuk Rebung Motif. Int J INTI 22(Dec):41–47
6. Dolchinkov N (2022) Marking and cutting of non-metallic products with CO_2 Laser. J Phys: Conf Ser 2224(1):012028
7. Štěpánková M, Wiener J, Dembický J (2010) Impact of laser thermal stress on cotton fabric. Fibres Text East Eur 80(3):70–73
8. Kan CW, Yuen CWM, Cheng CW (2010) Technical study of the effect of CO_2 laser surface engraving on the colour properties of denim fabric. Color Technol 126(6):365–371
9. Kan CW (2014) Colour fading effect of indigo-dyed cotton denim fabric by CO_2 laser. Fib Polym 15(2):426–429
10. Chow YL, Chan CK, Kan CW (2011) A study of physical modification on grey cotton by laser irradiation. Fib Polym 12(2):275
11. Juciene M, Urbelis VV, Juchneviciene Ž, Saceviciene V, Dobilaite V (2018) The influence of laser treatment and industrial washing on denim fabric tension properties. Int J Cloth Sci Technol 30(4):588–596. https://doi.org/10.1108/IJCST-03-2017-0032
12. Shamsuzzaman M, Abdul Awal Z, Das D (2021) Impact of laser intensities at various DPI and pixel time on the properties of denim garments. J Adv Res Mater Sci 77(1):1–13. https://doi.org/10.37934/arms.77.1.113
13. Yuan GX, Jiang SX, Newton E, Fan JT, Au WM (2012) Application of laser treatment for fashion design. J Text Inst 103(1):48–54
14. Yuan GX, Jiang SX, Newton E, Au WM (2013) Application of laser engraving for sustainable fashion design. Res J Textile Apparel

15. Yuan G (2018) Laser enhanced sustainable surface treatment for textile design on wool and polyester blended fabric
16. Guoxiang Y et al (2014) 8 technologies and design embellishment of aluminum foil laminated denim fabric using laser engraving technology. In: 9th International Shibori symposium 2014, pp 239–243
17. Bahtiyari MI (2011) Laser modification of polyamide fabrics. Opt Laser Technol 43(1):114–118
18. Shahidi S, Moazzenchi B, Ghoranneviss M (2013) Improving the dyeability of polypropylene fabrics using laser technology. J Text Inst 104(10):1113–1117
19. Akiwowo K, Kane F, Tyrer J, Weaver G, Filarowski A (2014) Digital laser-dyeing for polyester fabrics. J Textile Des Res Pract 2(2):133–151
20. Morgan L, Tyrer J, Kane F, Shen J (2014) Laser enhanced dyeing of wool for textile design, re-thinking text. Surfaces, Univ. Huddersfield, UK, pp 26–27
21. Montazer M, Taheri SJ, Harifi T (2012) Effect of laser CO_2 irradiation on various properties of polyester fabric: focus on dyeing. J Appl Polym Sci 124(1):342–348
22. Nourbakhsh S, Ashjaran A (2012) Laser treatment of cotton fabric for durable antibacterial properties of silver nanoparticles. Materials 5(7):1247–1257
23. Nourbakhsh S, Ebrahimi I, Valipour P (2011) Laser treatment of the wool fabric for felting shrinkage control. Fib Polym 12(4):521–527
24. Korzeniewska E, Gocławski J, Sekulska-Nalewajko J, Walczak M, Wilbik-Hałgas B (2020) Changes in fabric surface pilling under laser ablation. Sensors 20(20):5832

Numerical Modeling and Simulation of Micromachining of Biomedical Materials Using Nd: YAG Millisecond Pulse Laser

Brijesh K. Singh, Sajan Kapil, and Shrikrishna Nandkishor Joshi

1 Introduction

The term laser stands for "Light Amplification by Stimulated Emission of Radiation." The World's first laser was confirmed by Maiman using a Ruby Crystal in 1960. The light rays considered are of the same wavelength, i.e., monochromatic, highly collimated, and coherent. Laser beam machining is a non-conventional subtractive manufacturing process that uses thermal energy to remove metallic and non-metallic surfaces. Laser micromachining has gained much research attention due to its precision machining quality and efficiency. This machining is suitable for all materials such as metals, non-metals, conductive, non-conductive, and hard-to-cut materials. This machining process eliminates the tool wear problem, abrasion, and chattering problems due to its non-contact nature [1]. Due to its coherency and monochromatic nature, the concentrated laser energy is intended to melt higher-strength materials within seconds. When the laser beam transmits through the workpiece, it gets absorbed by the photon of the workpiece. Due to this phenomenon, the temperature of the workpiece raised rapidly to remove the material through melting and vaporization [2]. Thus the laser process is desired over other unconventional machining owing to its high ablation rate and excellent surface finish.

Titanium and its alloys are widely used materials in aerospace, and automobile industries due to their lightweight and attractive mechanical and physical properties. The biocompatible nature and good corrosion resistance make it suitable for biomedical applications, such as dental implants, hip prostheses, and knee joints [3, 4]. However, titanium and its alloy machining are challenging due to its low thermal conductivity, resulting in excessive heat generation at the tool and workpiece interface during the machining process [5]. Due to localized heating and the non-contact

B. K. Singh · S. Kapil · S. N. Joshi (✉)
Department of Mechanical Engineering, Indian Institute of Technology Guwahati, Guwahati, Assam 781039, India
e-mail: snj@iitg.ac.in

S. N. Joshi et al. (eds.), *Low Cost Manufacturing Technologies*,
https://doi.org/10.1007/978-981-19-8452-5_13

nature of laser beam machining, it would be the ideal processing technique for titanium alloys. But to avoid the defects such as heat affected zone (HAZ), micro-cracks, spatter formation, and bulges, using long laser pulses during processing are a big challenge for the researchers. Thus, systematic investigation using a finite element model and experimental methods is required to minimize these defects and establish laser micromachining in industries.

Thus it is crucial to consider these defects during laser micromachining of the components. This process consists of processes like cutting, drilling, marking, etc. The main challenges in these processes are getting a good aspect ratio during the drilling operation and sufficient materials removal during the cutting and marking operation. Researchers have reported numerical and experimental methods for micromachining of materials. Heng et al. used a low-power CO_2 laser to get the microchannel on the polycarbonate materials. They said the rough surface achieved an excellent aspect ratio with direct laser writing [6]. Zhang et al. [7] also studied the materials removal rate using an ultrafast laser on a silicon wafer. They experimentally and mathematically reported the effects of processing parameters such as scanning speed, laser frequency, and step distance on the ablation depth of machined surfaces. The surface quality of the laser machined a transparent material lithium niobate ($LiNbO_3$) for the medical device application was analyzed by Al-Shibaany et al. [8]. Various laser micromachining processes, such as cutting, drilling, and milling, were reviewed by Farasati et al. [9]. He reported the quality of cut and dimensional accuracy at various machining parameters, viz., laser power, gas pressure, scanning speed, and pulse duration for hard-to-cut material. Sahu et al. also studied the hybrid method to achieve a higher depth on C_p-titanium [10]. Achieving a good aspect ratio in micromachining requires more passes, leading to metallurgical effects on surface formation. So, from above literature indicates that there should be good combination of processing parameters to get better quality of results in terms of metallurgical and geometrical effects. More experimental increases the cost of manufacturing. Thus, numerical modeling needed to get the good combination of processing parameters for optimized results.

Numerical modeling is a powerful tool to predict accurate results during the laser micromachining of materials such as micro holes, microchannel, micro-groove, and micro-slots. Sarma et al. [11] numerically and experimentally investigated the geometrical size of the microchannel formed on polycarbonate using laser-induced plasma-assisted ablation (LIPAA) and predicted the ablated depth and width. Liu et al. [12] developed a 3D numerical model to predict width and depth to improve the tribological behavior of the tool materials. They also studied the processing parameters on the metallurgical effects such as heat affected zone (HAZ) and surface roughness. Remival of recast layer formation during laser machining by hybrid technique of laser machining and electrochemical machining is reported by Sun et al. [13]. However a comprehensive numerical model for micromachining is still missing in the current literature for optimized value of width and depth with minimum defects.

Scarce literature has been reported on using long laser pulses for micromachining of metallic materials. Thus in this work numerical modeling has been carried out to study the laser processing on biomaterials (Ti–6Al–4V) using Nd: YAG millisecond pulse laser. This model explores the temperature variation along axial and longitudinal directions with time. The developed numerical model is useful for metallic materials to predict the higher aspect ratio with minimum defects in single scan using long laser pulses.

2 Numerical Modeling and Boundary Conditions

The present study has developed a two-dimensional nonlinear transient finite element model of laser micromachining of Ti–6Al–4V having composition listed in Table 1. A temperature profile has been generated for moving heat sources by solving the finite element model considering temperature-dependent material properties of the titanium alloy (Ti–6Al–4V) listed in Table 2. The physical properties of the Ti–6Al–4V alloys have been presented in Table 3. The machining width and the machining depth have also been determined from the model.

2.1 Assumptions

Due to the complex phenomenon of the laser-matter interaction, certain assumptions have been considered to make the present model realistic. The assumptions are discussed as follows:

- Two dimensional (2D) model of the material has been considered.
- Laser source is considered to be a Gaussian profile [18].
- The material is considered isotropic and homogeneous.

Table 1 Chemical composition of Ti–6Al–4V alloy [14]

Element	C	N	O	Cr	Al	V	Fe	Ti
% Composition	0–0.1	0–0.5	0–0.2	0.021	5.5–6.75	3.5–4.5	0–0.4	Balanced

Table 2 Temperature-dependent thermal properties of Ti–6Al–4V alloy [15]

S. No.	Temperature (°C)	Specific heat capacity c_p (J/kgK)	Thermal conductivity k (W/mK)
1	27	403	6.5
2	300	505	7
3	400	520	9
4	500	530	11
5	600	545	12
6	800	620	17

Table 3 Physical properties of Ti–6Al–4V alloy [16, 17]

	Properties	Value
Ti–6Al–4V	Density, ρ (kg/m^3)	4430
	Absorptivity	0.48
	Melting point, T_{m} (K)	1941
	Evaporation temperature, T_v (K)	3560
	Emissivity, ε	0.708
	Latent heat of melting (J/kg)	4.2×10^5

- It is considered that material removal will occur when the temperature reaches above the melting point of the workpiece material.
- Reference temperature is 300 K.
- Plasma generation and multiple reflections of the laser beam were ignored.

2.2 Governing Equation

The two-dimensional partial differential equation governs heat flow in the workpiece material in Cartesian coordinates

$$\frac{\partial}{\partial x}\left(k\frac{\partial T}{\partial x}\right) + \frac{\partial}{\partial y}\left(k\frac{\partial T}{\partial y}\right) = \rho c_p \frac{\partial T}{\partial t} \tag{1}$$

where T is the temperature, k is the thermal conductivity (W/mK), ρ is the density of materials (kg/m^3), t is the time (s), and c_p is the specific heat of the workpiece materials (J/kg K).

2.3 Boundary Conditions

The boundary condition consists of a Gaussian heat source, convection, and radiation on the thermally affected boundaries AB of the workpiece material as represented in Fig. 1. The other sides BC, AD, and CD, were considered insulated, the boundary condition of the described model can be expressed mathematically as follows:

$$\text{Boundary AB}\quad k\left(\frac{\partial T}{\partial x}\right) + k\left(\frac{\partial T}{\partial y}\right) = -h_t(T - T_{\mathrm{ref}}) - \varepsilon\sigma\left(T^4 - T_{\mathrm{ref}}^4\right) \tag{2}$$

$$\text{Boundary BC, AD and CD}\quad k\left(\frac{\partial T}{\partial n}\right) = 0 \tag{3}$$

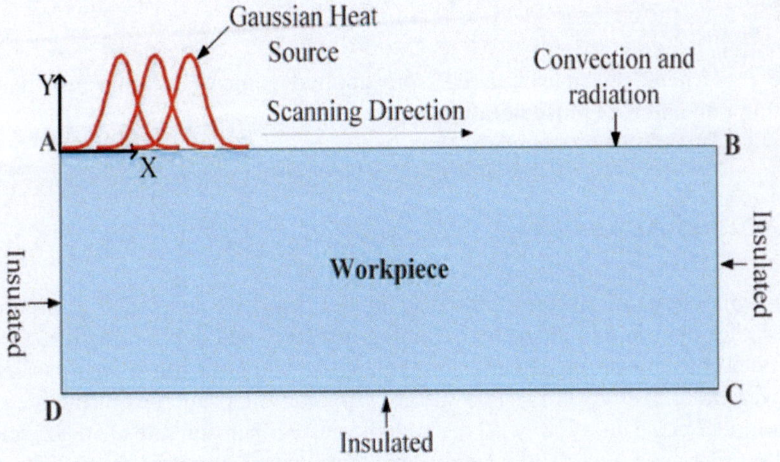

Fig. 1 Geometric representation of the model and boundary conditions

where h_t is the convective heat transfer coefficient, T is the surface temperature, ε is the emissivity of the workpiece material, and σ is the Stefan Boltzmann constant ($5.67 \times 10^{-8} \text{Wm}^{-2}\ \text{K}^{-4}$), T_{ref} reference temperature at $t = 0$, time, and n is the normal surface of the irradiated heat zone.

The initial condition of the workpiece material is taken to be at an ambient temperature of 300 K at a time $t = 0$

$$T_{\text{ref}} = 300\,\text{K} \tag{4}$$

2.4 *Heat Flux Model*

The irradiated heat source top surface of the workpiece is considered a Gaussian distribution, represented as [16]

$$Q = \frac{2A_s P_a}{\pi r_s^2} \exp\left(\frac{-2x^2}{r_s^2}\right) \tag{5}$$

$$f(t) = \left.\begin{matrix} 1\ t < t_{\text{p}} \\ 0\ t > t_{\text{p}} \end{matrix}\right\} \tag{6}$$

where Q is the applied heat flux on the top surface of the workpiece (W/m^2), A_s is the material's absorptivity, t_{p} is pulse duration P_{a} is the average power, r_s is the laser beam diameter considered to be 0.2 mm during the entire simulation.

Average power can be calculated from the Eq. (7)

$$P_a = P_k t_p f \tag{7}$$

where P_k is peak power calculated from applied voltage and current, f is pulse repletion rate and t_p is pulse duration.

3 Solution Approach

A 2D finite element transient numerical model has been developed to predict the temperature distribution along the scanning direction and depth direction. Reasonable material removal rate depends on the appropriate laser processing parameters. However, more heat generation on surface can alter the properties of the materials. In this model pulse duration is 1.0 ms, and the pulse repetition rate is 50 Hz, selected based on reviewed paper and scanning velocity consider to be 4 mm/s. The entire model was coarse meshed, and the heated and nearby region was discretized with uniform fine mesh of 10 μm to minimize the simulation time, as shown in Fig. 2. The triangular elements type is used for the whole model. To solve this presented model time dependent, MUMPS solver used with relative tolerance of 0.005.

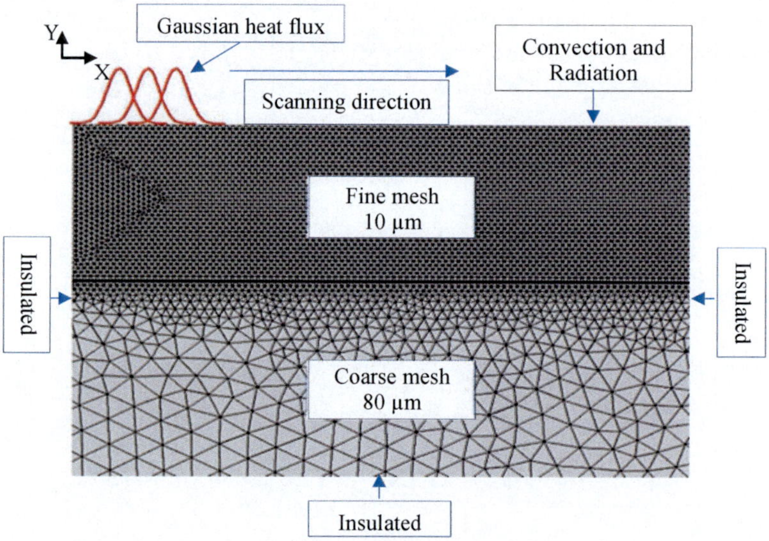

Fig. 2 Mesh process continuum

4 Results and Discussion

4.1 Transient Thermal Analysis of Laser Machining Process

Transient thermal analysis of the laser micromachining process was carried out based on the two-dimensional (2D) temperature distribution shown in Fig. 3. the variation of temperature along the scanning direction is represented. It shows the isotherm contour for the workpiece materials along the moving path for 1 ms pulse duration and 50 Hz pulse repetition rate at different times 0.125 s, 0.5 s, and 1.0 s. The temperature of the workpiece increases along the scanning direction. This is because the moving heat preheats the surface. It can also be seen in Fig. 4. As time passes, the temperature of the workpiece materials increases and attains a specific maximum temperature, also similar trend observed by Sarma et al. during microchannel fabrication on polycarbonate materials using laser-induced plasma-assisted ablation (LIPAA). For the laser machining to get the uniformity of the surface roughness, the variation in the temperature limit should be minimum to get the uniform geometry such as depth and width. Increasing temperature leads to the formation of the melt pool and further comes out from the surface, known as ablation of materials.

During the long laser pulse interaction with the material, when the surface temperature exceeds creates a pressure difference (inside melt and atmospheric pressure) which becomes sufficient to expel materials from the surface, leaving a cavity on the surface [19]. Sharma et al. also analyzed the melt pool behavior during materials processing using long laser pulses [20].

4.2 Prediction of Depth and Width During the Machining Process

The effect of the processing parameters 1 ms pulse duration, and 50 Hz pulse repetition rate was studied on machining geometry, such as machining depth and width in single passes of laser scanning. Above Fig. 5a, b represent the prediction of machining width and depth, respectively. For the prediction machining width, a temperature profile plotted along the scanning direction at time 0.25 s. It was considered that when the surface temperature of workpiece will exceed the melting point materials will come out due to pressure difference. The predicted width is 290–300 μm as shown in Fig. 5a. For the prediction of removing depth during the machining process, a line considers along the depth direction at same time 0.25 s. Temperature profile plotted along this line which shows the temperature decreases along the depth direction which happens due to the Beer lambert principle (due to decrement in intensity) also reported in previous literature. A constant melting point line of workpiece materials plotted in the graph and at the point where it intersects the temperature profile gives the to predict depth during the machining process. The predicted machining depth is

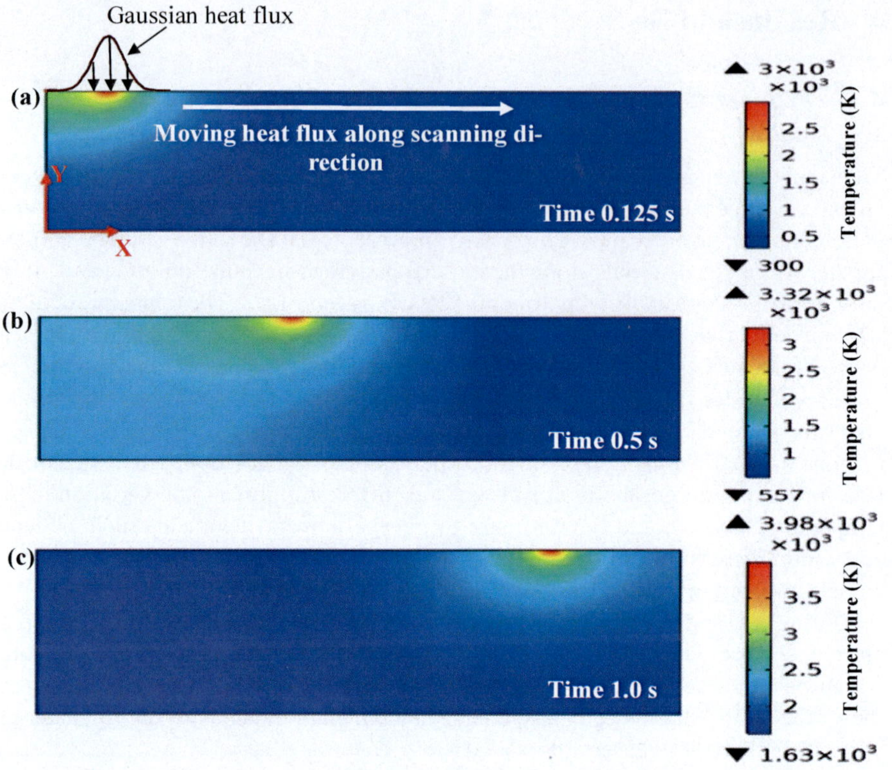

Fig. 3 Temperature distribution on the workpiece at **a** 0.125 s, **b** 0.5 s, and **c** 1.0 s along the scanning direction at 1.0 ms pulse duration and 50 Hz pulse repetition rate

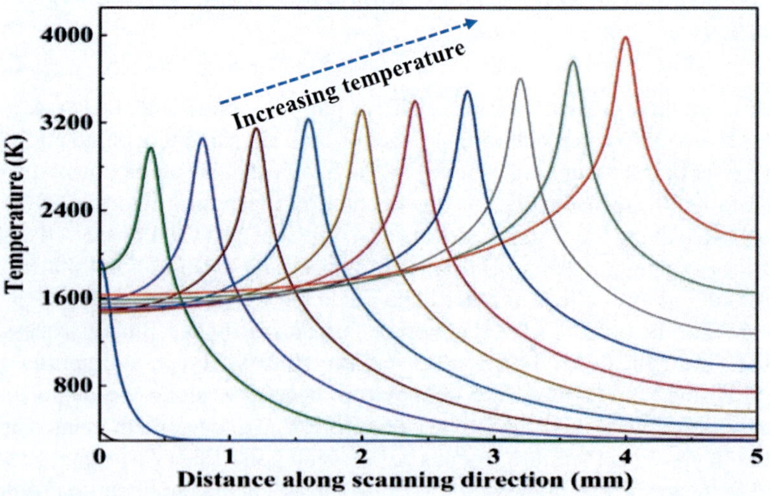

Fig. 4 Temperature profile along moving direction

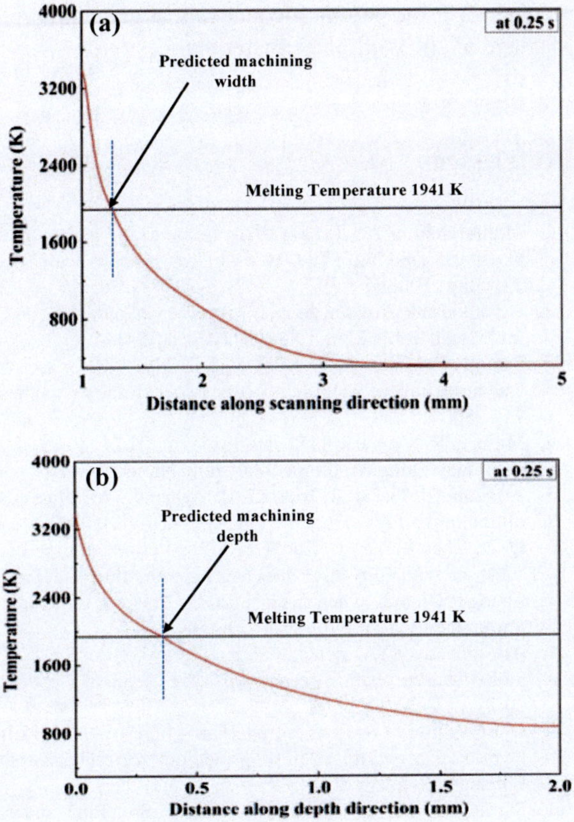

Fig. 5 **a** Temperature distribution along scanning direction **b** depth direction at 1 ms pulse duration and 50 Hz pulse repetition rate at time of 0.25 s

around 310–320 μm. It was expected that during the machining process there will be less detrimental effects such as micro-cracks, spatter on the workpiece.

5 Conclusion

The present work presents a 2D finite element numerical model of laser micromachining of a biomedical material, i.e., titanium alloy (Ti–6Al–4V) with pulse duration of 1 ms and 50 Hz pulse repetition rate. Machining depth and width were predicted based on the temperature variation profile along the scanning direction and depth direction, respectively. Materials removal is considered after exceeding the melting point of the workpiece material. The uniform melt formation along scanning direction and depth direction forms uniform width and depth during the machining, which is desirable for a good surface finish. The presented model predicts a machining width of 290–300 μm and machining depth of 310–320 μm. This model can further be extended for the parametric analysis of machining geometry and thermally affected damage on the workpiece materials.

References

1. Muthuramalingam T et al (2020) Influence of process parameters on dimensional accuracy of machined Titanium (Ti–6Al–4V) alloy in laser beam machining process. Opt Laser Technol 132(July):106494
2. Parandoush P, Hossain A (2014) A review of modeling and simulation of laser beam machining. Int J Mach Tools Manuf 85(Oct 2014):135–145
3. Saternus Z, Piekarska W, Kubiak M, Domański T, Goszczyńska-Króliszewska D (2019) Numerical modeling of cutting process of steel sheets using a laser beam. MATEC Web Conf. 254:08004
4. Miller PR, Aggarwal R, Doraiswamy A, Lin YJ, Lee YS, Narayan RJ (2009) Laser micromachining for biomedical applications. JoM 61(9):35–40
5. Arrazola PJ, Garay A, Iriarte LM, Armendia M, Marya S, Le Maître F (2009) Machinability of titanium alloys (Ti6Al4V and Ti555.3). J Mater Process Technol 209(5):2223–2230
6. Qi H, Chen T, Yao L, Zuo T (2009) Micromachining of microchannel on the polycarbonate substrate with CO_2 laser direct-writing ablation. Opt Lasers Eng 47(5):594–598
7. Zhang P, Chen L, Chen J, Tu Y (2017) Material removal effect of microchannel processing by femtosecond laser. Opt Lasers Eng 98:69–75
8. Al-Shibaany ZYA, Penchev P, Hedley J, Dimov S (2020) Laser micromachining of Lithium niobate-based resonant sensors towards medical devices applications. Sensors (Switzerland) 20(8)
9. Optimization of laser micromachining of Ti–6Al–4V | Elsevier Enhanced Reader
10. Hybrid laser and micro milling methods for higher depth microchannel fabrication_Elsevier Enhanced Reader.pdf
11. Sarma U, Joshi SN (2020) Numerical modelling and simulation of microchannel fabrication on polycarbonate using Laser-Induced Plasma Assisted Ablation (LIPAA). Optik (Stuttg) 223(Aug):165379
12. Liu Y, Liu X, Li T, Tian Y (2021) Numerical modelling and experimental study on pulsed laser surface texturing on cemented carbides. Int J Adv Manuf Technol 114(9–10):3137–3145
13. Sun A, Chang Y, Liu H (2019) Numerical simulation of laser drilling and electrochemical machining of metal micro-hole. Optik (Stuttg) 181(Nov 2018):92–98
14. Pramanik A, Basak AK (2014) Sustainability in wire electrical discharge machining of titanium alloy: understanding wire rupture. J Clean Prod 198(Pramanik 2014):472–479
15. Yilbas BS, Akhtar SS, Karatas C (2012) Laser hole cutting into Ti–6Al–4V alloy and thermal stress analysis. Int J Adv Manuf Technol 59(9–12):997–1008
16. Xie BC, Wang YK, Wang ZL, Zhao WS (2011) Numerical simulation of titanium alloy machining in electric discharge machining process. Trans Nonferr Met Soc China (English Ed) 21(SUPPL 2):s434–s439
17. Kiran Kumar K, Samuel GL, Shunmugam MS (2019) Theoretical and experimental investigations of ultra-short pulse laser interaction on Ti6Al4V alloy. J Mater Process Technol 263(May 2018):266–275
18. Ren N, Jiang L, Liu D, Lv L, Wang Q (2015) Comparison of the simulation and experimental of hole characteristics during nanosecond-pulsed laser drilling of thin titanium sheets. Int J Adv Manuf Technol 76(5–8):735–743
19. Wagner RE (1974) Laser drilling mechanics. J Appl Phys 45(10):4631–4637
20. Sharma S, Mandal V, Ramakrishna SA, Ramkumar J (2018) Numerical simulation of melt hydrodynamics induced hole blockage in Quasi-CW fiber laser micro-drilling of TiAl6V4. J Mater Process Technol 262(May):131–148

A State-of-the-Art Review on Surface Modification Techniques in Electric Discharge Machining

Binoy Kumar Baroi, Tapas Debnath, Jagadish, and Promod Kumar Patowari

1 Introduction

This paper describes the literature of electrical discharge machining (EDM) process for the application of surface modification. EDMed surface layer contains a white layer that crystallizes for molten liquid metal, which is cooled at a higher rate. After examining the white layer, some of the tool material was accumulated on the work surface by the re-solidification process [1–3]. Thus, the metallurgical structure of the surface was changed. This re-solidification phenomenon was suggested by many researchers for surface modification purpose of the material. This way, the idea of surface modification was developed gradually. Many researchers used powder metallurgical (P/M) tool for having a desired coating composition on the work surface using electric discharge coating (EDC). P/M tool provided a uniform modified surface. The tool material transferred from the tool to the workpiece more effectively at reverse polarity [4]. Ahmed et al. [5] proposed that straight tool polarity was of utmost suitable if the sole aim was to lessen the tool wear rate (TWR). But, for minimizing the

B. K. Baroi (✉) · P. K. Patowari
Department of Mechanical Engineering, National Institute of Technology Silchar, Silchar, Assam 788010, India
e-mail: baroibinoy@gmail.com

P. K. Patowari
e-mail: ppatowari@mech.nits.ac.in

T. Debnath
Department of Mechanical Engineering, Techno India University, Kolkata, West Bengal 700101, India
e-mail: debnath.tapas@yahoo.com

Jagadish
Department of Mechanical Engineering, National Institute of Technology Raipur, Raipur, Chhattisgarh 492010, India
e-mail: jagadish.mech@nitrr.ac.in

S. N. Joshi et al. (eds.), *Low Cost Manufacturing Technologies*,
https://doi.org/10.1007/978-981-19-8452-5_14

overcut, reverse polarity should be used. Kumar et al. [6] reported that machinability and surface modification largely depends on the dielectric type. Powder particles mixed in the dielectric reacted with the surface of the substrate and created a coating over it in case of powder mixed EDM (PMEDM). Bhattacharya et al. [7] studied EDM characteristics with different workpieces (HCHCr, H11, AISI 1045 steel and hot die steel), tools (Gr, W–Cu and brass), and powder particles (Si, graphite and W) mixed with kerosene and EDM oil, respectively. According to observations, W–Cu tool and W powder particles produced better microhardness while brass tool and W powder particles produced decent surface quality. A greater cooling effect was produced by using kerosene as the dielectric, and the surface finish was improved on using EDM oil. Amorim et al. [8] used Mo powder for machining of AISI H13 in EDM. The modified surface contained Fe–Mo, MoxC and Mo, which enhanced the surface characteristics. Jabbaripour et al. [9] investigated the surface morphology with variations in control parameters. With the rise in current (I_p), larger crack density and deeper cracks were observed. Also, with a growth in pulse on time (T_{on}), the microhardness increased. Bhaumik and Maity [10] used Cu, brass and zinc tools to ED machine Ti–5Al–2.5Sn in EDM. It was observed that higher MRR was achieved using brass and zinc tools followed by the Cu tool. TWR was less for Cu followed by brass and zinc. Cu tool produced surface with less crack density and high surface finish.

The present work covers machinability and surface modification using EDM in detail. Other than machining, the surface of the workpiece can be modified in EDM. Various things need to be considered for improving the surface quality, be it in terms of surface finish or surface hardness. P/M metallurgy and PMEDM are the essential techniques to upgrade surface quality. But, proper selection of the process parameters, dielectric type and the tool-workpiece combination is also equally important.

2 Surface Modification Using EDM

The EDM method is mostly used to improve surface quality in addition to machining. Better surface characteristics like microhardness, wear and corrosion resistance can be achieved using EDM. The following strategies enhance these qualities.

2.1 Surface Modification By Electric Discharge Coating

Researchers have verified that the electrode wear is higher in the case of P/M green compact or sintering tool than a solid electrode. It is owing to low thermal conductivity and low bonding among the powder particle of P/M tool and P/M semi-sintered tool. It encourages the use of P/M tool for surface improvement by moving the considerable amount of tool material on the substrate. Samuel and Philip [11] performed a comparative analysis between P/M tool and conventional electrode. They reported that P/M tools were more prone to variations in T_{on} and I_p. Performance measures were different compared to the conventional electrode. They noticed that P/M electrode could result in material addition instead of material removal under certain operating conditions from the workpiece. Pantelis et al. [12] tried to increase the surface hardness of tool steel by sintered 70Fe–30WC electrode through the development of a compound layer on the workpiece comprising of anti-wear elements similar to carbides. This process could increase the mechanical factors influencing the substrate and the functional life under various working conditions. Hardness also could be improved by EDM process. Wang et al. [13] recommended the procedure of surface modification by EDC for cutting tools and moulds. They performed experiments on carbon steel with TiC P/M electrode (more than 51% weight percentage) and resulted in a more than 3 times increase in hardness. It was due to the movement of elements from tool to substrate. The formation of metal carbide in the white layer aided in to increase in hardness.

Gangadhar et al. [14] improved the surface characteristics of mild steel using a bronze (90%Cu–10%Sn) compact tool in reverse polarity. The result showed that Cu_3Sn_5 and CuSn were the primary constituents in the substrate. They reported the possibility to change the physicochemical and metallurgical properties of the surface by appropriate changes in the constituents and elements of the P/M tool. Mohri et al. [15] modified the topography of Al and carbon steel using tools of Al, Cu, Ti and WC in hydrocarbon oil. They reported that the EDMed topography was found to be corrosion and wear resistant with fewer cracks. Thus, surface integrity can be improved by changing the metallurgical structure of the work material. H13 tool steel surface was modified by WC/Co electrode. The modified layer resulted less microcracks, and surface hardness increased from 640 to 1319 HK [16]. Open-gap voltage showed little effect on microhardness. Shunmugam et al. [17] applied WC powder compact electrodes of 60%Fe and 40%WC to make mild steel wear resistant. An enhancement in abrasive wear resistance was observed by 25–60%. Figure 1 depicts the microhardness achieved with different composition of W–Cu. 60:40 wt% composition offered the highest microhardness among all the compositions [18].

Long durability is a desired characteristic of a functional product. It depends upon the wear resistance, corrosion resistance, hardness and surface characteristics. Singh and Kumar [19] used a P/M composite Cu–Mn tool to process hot die steel. The highest microhardness achieved was 1191.7 HV. The microhardness improved by 93.7% compared to the parent material. The surface roughness (SR) was observed to be 3.11 μm with no microcracks. Kruth [20] proposed that reverse polarity favoured high tool wear while depositing Al on steel and TiC using Al and Ti–Al P/M tools. Ho et al. [21] performed surface modification of Ti–6Al–4V with both solid and P/M Cu tool using commercially available water-based dielectric fluid. Recast layer thickness (RLT) in between 4 and 11 μm was achieved with rougher and thicker layer at straight polarity. The RLT possessed hardness up to 1100HK, compared to the hardness of bulk material 365 HK. Thus, the desired properties of substrate could be improved by surface modification process. Patowari et al. [22] improved the surface of C-40 steel using an electrode of WC–Cu. They detected just a small

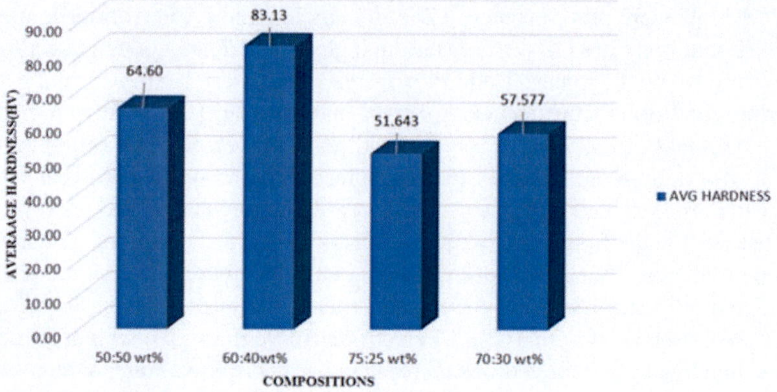

Fig. 1 Hardness of coated surface at different compositions [11]

number of microcracks, and the hardness rose from 200 to 220 HV to 1200 to 1632 HV.

Hwang et al. [23] created a film of TiC on the surface of Ni workpiece by EDC utilizing electrodes with many layers consisted of Ti and Gr. They reported that the graphite layer enhanced the carbon concentration, and TiC improved surface hardness. Additionally, TiC lowered the coated layer's surface irregularity and lowered the probability of microcracks formation. Patowari et al. [24] applied W–Cu P/M sintered tool to improve the topography of C-40 steel. The deposited layer's thickness ranged from 25 to 739 μm, having the particles of W and Cu owing to the formation of TiC. They found microhardness in the range of 9.81–12.75 GPa.

In the P/M technique of EDM, material is deposited from tool on the work surface rather than being removed from the workpiece. W, WC, Ti, TiC are the most common powders that are used in making P/M tool. This is because these powders are electrically conductive as well as these can provide a very hard and wear resistant layer on the surface with good surface integrity. To improve the electrical conductivity of the tool electrode, Cu powder is generally utilised in combination with P/M powders.

2.2 *Surface Modification By Powder Mixed EDM*

PMEDM is one of the inventive ideas to enhance the process capabilities for a better performance measure, mainly for surface properties. The addition of micron-sized metal particles to the dielectric fluid expands the possibilities for workpiece surface modification. Ti, Si, Fe, Co, W and Ni powders with sizes ranging from 1 to 100 μm are frequently employed [25–27]. The ideal particle size is less than 50 μm to avoid the spark gap problem [28]. The addition of powder particles improved the breakdown properties of the dielectric by causing an abnormal rise in the spark gap's electric field. The provided gap voltage results in the bridging effect at the electrode gap by forming positive and negative charges on the upper and lower part of the abrasive particles [21]. Electric discharge thus accumulated to trigger discharge between the two successive particles of powder, which results in a sequence of discharge [29, 30]. Breakdown happened where the I_p density is the maximum and when the field's density exceeds the dielectric's breakdown resistance. This phenomenon improved the spark intensity, which results in faster erosion. The abrasive elements enlarged the spark gap, thereby reducing the gap voltage. Due to the creation of large-diameter craters on the substrate, the expanded discharge channel also upsurges the discharge heat route and decreases the discharge density, thereby improving the surface finish[31, 32]. Figure 2 depicts the relationship between powder concentration and duty factor with microhardness. It was found that microhardness incremented with increment in duty factor. It was also clear that microhardness enhanced with the enhancement in powder concentration [33].

Al, Si and Gr powders were added to the dielectric to increase the surface polish and create a mirror finish surface[32]. SiC powder increased the machining gap and machining rate whereas superior surface finish was attained with Al powder added dielectric [33]. Wu et al. [34] studied how adding Al powder and surfactant to the dielectric had an impact on the surface. They found best effect at concentration of

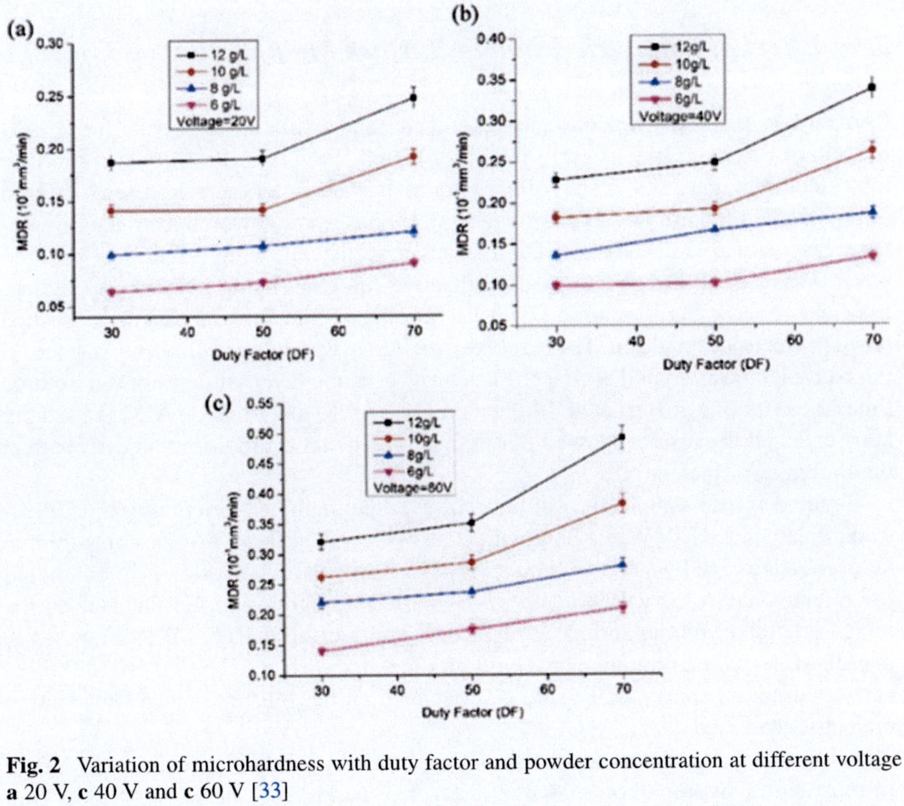

Fig. 2 Variation of microhardness with duty factor and powder concentration at different voltage **a** 20 V, **c** 40 V and **c** 60 V [33]

Al powder 0.1 g/L and surfactant 0.25 g/L, in the dielectric. The SR was reduced up to 60% for PMEDM compared to pure dielectric. Microhardness increases, and the microcracks decreases due to the existence of additives in the dielectric [35]. Luo [36] reported an improvement in discharge transitivity and machining stability during EDM due to a decrease in arcing frequency as a result of evenly dispersed debris in the gap. Ojha et al. [37] studied the impact of Ni micro-powders by suspending in the dielectric. They reported that the layer formed after machining resulted decent surface finish and increased the hardness and strength of the work material to appreciable amount. Kumar and Batra [38] modified the surface of H13 die steel material using W powder added dielectric. They found a maximum quantity of 3.25% W on the topography of H13 die steel.

PMEDM mostly used by the researchers to enhance the machinability of the process. Electrically conductive powders can impove the MRR and surface finish by transferring the spark energy from the electrode to the workpiece in a uniform way. As a result, material from the workpiece remove uniformly all over the surface. Additionally, the workpiece's surface develops a coating as a result of the powder components' reaction with the workpiece material. The work surface's properties are improved by these coatings.

2.3 *Effect of Dielectric Fluid in Surface Modification*

Comparison of various dielectric, used kerosene, and distilled water for surface modification of Haynes 230 was performed by Bai and Koo [39] with Mo–Al composite tool electrode. Distilled water offered more hardness, whereas kerosene offered improved SF, thicker layer, better surface morphology and lowered the rate of oxidation. They stated that kerosene had a stronger surface alloying effect than purified water. Wang et al. [13] performed surface modification using kerosene as dielectric fluid. A very hard-ceramic layer was applied to the substrate using a crushed powder electrode made of Ti or another metal. It was noticed that the surface of the workpiece was coated with a compact TiC ceramic layer which resulted in three times more hardness. Beri et al. [40] tested Cu–W P/M electrode on AISI D2 steel in kerosene, and the results showed that it performs better on multiple objectives than the traditional Cu electrode.

Figure 3 informs about the variation of average diameter, machining time, electrode wear and RLT. Water-based dielectric shows the best results compared to kerosene, deionized water and emulsion [41]. Kunieda and Yoshida [42] compared the effectiveness of dry EDM and oil-based EDM. They reported that due to the increased crater volume and more recurrent supply of discharge, MRR increases in dry EDM. Metal carbide was observed on the work surface when oil-based dielectric was used, whereas metal oxide was detected on the substrate when water-based dielectric was used.

Selection of the dielectric is also an important task while machining in EDM. In most cases, hydrocarbon oil offers a layer of carbide on the work surface. But,

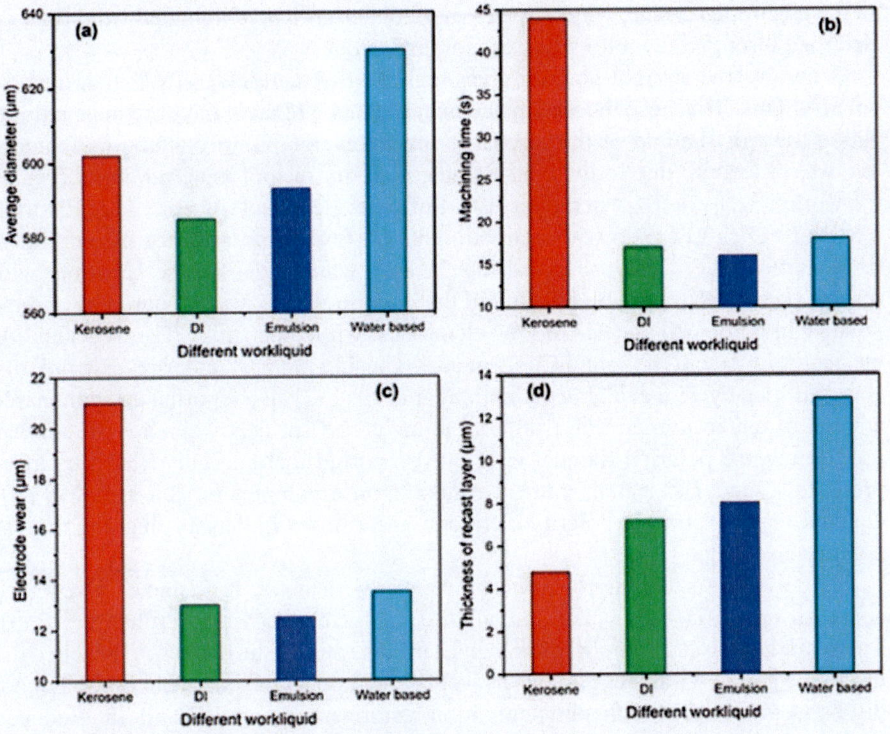

Fig. 3 Results of different dielectric fluids with the same discharge parameters [34]

many researchers avoid the use of hydrocarbon oil as it pollutes the surrounding. Water-based dielectrics are used as a substitute of the hydrocarbon oil. Water-based dielectrics form a oxide layer on the surface. Various types inert gases are also used by the researchers to make the process green. These gases act as coolant during machining. Also, the flushing action is excellent in dry EDM.

2.4 *Surface Modification By Controlling the Parameters*

Discharge voltage is the leading parameter that effects the surface treatment. This is directly associated to discharge gap and strength of the dielectric [43]. Higher spark gap and better flushing conditions are produced by the higher voltage setting. Increased open-circuit voltage increased the strength of the electric field, which caused electrodes to wear and eventually embed tool material on the work surface [44]. The input I_p in machining determines the amount of energy used [45]. The I_p is administered by surface area to be cut. High I_p and longer pulse duration lead to high TWR because they increase the heat energy input and increase the thermal energy on the electrode surface, which causes more material to be eliminated from the tool electrode through the melting and evaporation process.

Generally, the straight polarity gives higher MRR and lower TWR than reverse polarity. This is because the workpiece material had a relative increase in heat dissipation towards the end of the discharge time. Reverse polarity alters the material removal procedure due to the considerable quantity of tool material depositing on the workpiece [46]. To observe the effect of polarities, Pantelis et al. [12] ED used a tool (70%Fe–30%WC) with both straight and reverse polarity for machining of 0.4% carbon steel. They observed that W rich white layer on the substrate with reverse polarity. Kruth et al. [20] found that high tool wear was favoured by reverse polarity in porous Al and Ti–Al P/M electrodes, while depositing TiC on Al and Al on steel. Movement of Cr and Cu from a Cr–Cu sintered tool under reverse polarity was informed by Tsai et al. [47]. Shunmugam et al. [17] also reported an appreciable amount of material migration from tool to machined surface, when a P/M tool was used at reverse polarity. Samuel et al. [48] reported that straight polarity is more prone to I_p and T_{on}, which influence the response parameters. Bai and Koo [39] claimed that the straight polarity electrode has a lower $\mathrm{I_p}$ density than the reverse polarity electrode.

Figure 4 reports that microhardness enhances with an rise in Ni and Cr percentage composition. I_p, T_{on} and voltage are all directly correlated with microhardness [49]. The spark gap plays an important role in EDM operation. The dielectric breached the discharge gap to eliminate gaseous as well as hard debris during machining. Higher discharge gap improved flushing condition resulting higher MRR and SR. Reduced gap length resulted in ignition delay. Wong et al. [50] found that Al powder offered both larger spark gap and MRR. Yan et al. [51] reported that Cr and Al mixture in the kerosene increased spark gap and reduced isolation. This helped in making the process stable, and the MRR was upgraded considerably. Moreover, due to the low gap between the tool electrode, the carbon particles and debris cannot be removed easily, which results an uneven spark.

Flushing is a beneficial process to eliminate debris from machining region. Researchers working on flushing pressure report that it influences the SR and TWR. It also acted as a coolant and removed debris particles from the machining area [52]. MRR was highly influenced by flushing [53, 54]. During roughing process, flushing affected the MRR and TWR [54]. Finding the ideal flushing velocity helps in reducing the impact of flushing velocity on RLT and crack density[55]. The dielectric can be provided up or down through tool electrode by means of jet flushing and vacuum flow [56]. Dielectric can enter from one side and exit from the other side owing to the relative movement of the substrate and electrode along with planetary tool movement[57]. Application of magnetic field and controlled forced vibration vacated debris successfully from the spark zone [58]. Flushing can be improved by using ultrasonic vibration on both the tool and the workpiece [59]. Ghoreishi and Atkinson [60] stated that the amalgamation of tool rotation and ultrasonic vibration rises the TWR and MRR. The vibration cum rotary motion improved the MRR by approximately 100% compared to rotary EDM, whereas the improvement is up to 35% compared to vibration EDM in semi-finish operation. Zhang et al. [61] performed

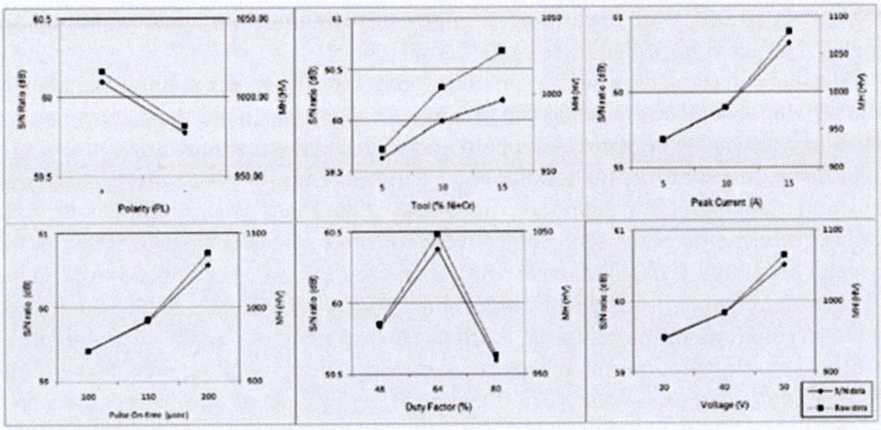

Fig. 4 Main effect plots for microhardness [42]

ultrasonic EDM in gas with a thin hollow tool. The outcome showed increased MRR with longer pulse length, higher open voltage discharge I_p, higher ultrasonic actuation and thinner tool pipe walls. Gunawan et al. [62] reported that the effect of flushing increases with the vibration of workpiece. The high frequency with high amplitude increases MRR. Xu et al. [63] introduced gaseous EDM with an ultrasonic vibration-assisted tool electrode. Experimental outcomes proved that MRR could be improved by introducing ultrasonic vibration.

The most frequently used process parameters are I_p, T_{on}, pulse off time (T_{off}), voltage, spark gap, polarity, powder concentration, flushing rate and dielectric depth. Process parameters with appropriate levels are essential to initiate the machining process. At very low parameter setting, machining does not occur. Again, at very high parameter setting, the machining occurs absurdly. At high energy, arcing occurs in place of sparking, which is not controllable. Also, at high-energy settings, the surface finish deteriorates. Moreover, the crack formation also increased at a high level of parameter setting.

2.5 *Parametric Optimization in Surface Modification*

Taguchi technique gives a powerful fractional factorial method instead of the full factorial experimental method. The Taguchi design of experiment (DOE) method is a strong and useful tool for creating processes that perform dependably and ideally under various circumstances. Simao et al. [16] performed statistical analysis applying L-8 fractional factorial design to modify the surface of H13 hot working tool steel by WC/Co tool and recognized the influence of significant input parameters to reduce the microcracks with increased surface hardness. Patowari et al. [24] applied L-16 orthogonal array of Taguchi DOE technique for surface modification. Overall evaluation criteria were used to get an optimum condition for smooth and regular material deposition, which increased microhardness.

Similarly, fuzzy logic was applied to optimize the variable input parameters considering several output responses [64, 65]. Recognizing this, Takagi and Sugeno's method was used to construct an input–output link in the manufacturing process by applying particular fuzzy modelling [66]. Puerts and Luis [67] used a factorial design of studies and regression analysis to model Ra of the EDM process. One of the most well-known methods for multi-objective optimization in the field of EDM is the genetic algorithm (GA). Deb et al. [68] optimized the processing conditions using the non-dominated sorting genetic algorithm (NSGA-II), and a set of non-dominated optimal solutions were identified. Artificial neural network (ANN), a paradigm of artificial intelligence, has many attractive properties for modelling complex systems containing many process data [69]. Tsai and Wang [70] developed several models of neural networks, including an adaptable neuro-fuzzy inference system (ANFIS).

Researchers used different optimization techniques in EDM to optimize the control variables so that all the response measures are improved at a particular parameter setting. At the optimum parameter, the machinability and surface characteristics such as hardness, wear resistance, corrosion resistance and surface finish are also improved. I_p, T_{on}, T_{off}, voltage, spark gap, polarity, powder concentration, flushing rate and depth of dielectric are the most common control variables that are optimized using various optimization techniques.

3 Future Scope

Though EDM is extensively used in various industry and research laboratories, it has some demerits that need to be addressed to increase its efficiency. In many case studies, MRR of EDM is much less than some conventional machining processes. The thickness of the coating in P/M technique is more than in the PMEDM process. Proper powder concentration particles in the dielectric must be selected to obtain more coating thickness. Most of the powder used in PMEDM has very low hardness value. Researchers should consider using hard and good alloying powders such as Ti, W, Cr, Mo and Co. The effect of size of the powder material on surface modification is also an important area for study. In most PMEDM studies, the most common parameters are I_p, T_{on} and powder concentration. But, the workpiece surface gets quenched during T_{off}. Hence, more studies should be carried out, taking T_{off}. Variation of surface hardness and surface finish with the powder particle size should be considered while experimenting. Other than EDM, surface modification can also be performed by various other techniques such as electroplating and spray coating. Hence, comparison should be done between these two techniques regarding quality of the workpiece and cost of machining.

4 Conclusion

The use of EDM in different industry sectors, such as tool rooms, die-making industry and general shop floors, is increasing every day due to its capacity to produce complex profiles on difficult-to-machine materials. The surface modification adds a new dimension to EDM due to the improvement of various mechanical properties. The following conclusions are drawn from the detailed literature survey.

1. Various powders are used in the EDC process to get the desired coating on the work surface. Generally, W, WC, Ti and Co powders are used in EDC process.
2. PMEDM is an effective technique used for machining and coating simultaneously. Powders increase the rate of material removal. It also smoothens the surface finish. Powders like Ti, TiC, B_4C, Si, Co, Ni and W improve the surface characteristics such as microhardness, wear and corrosion resistance.
3. In EDM, kerosene and hydrocarbon oil are the dielectrics most often utilized. Water and gas-based dielectrics are also used as a substitute of hydrocarbon oil and kerosene oil. The discharge energy is delivered to the work surface more efficiently when powder particles are present because they reduce the dielectric's insulating strength.
4. The most commonly used process parameters are I_p, T_{on}, T_{off}, voltage, spark gap, polarity, powder concentration, flushing rate and depth of dielectric. At high I_p and T_{on}, the workpiece's surface finish deteriorates. Also, SR and the number of cracks on the surface increased.
5. Optimization is an effective tool to decide the combination of control variables that provide the optimum MRR, TWR, SR and surface qualities such as microhardness, wear and corrosion resistance.

References

1. Zhao W, Meng Q, Wang Z (2002) The application of research on powder mixed EDM in rough machining. J Mater Process Technol 129(1–3):30–33. https://doi.org/10.1016/S0924-0136(02)00570-8
2. Furutani K, Sato H, Suzuki M (2009) Influence of electrical conditions on performance of electrical discharge machining with powder suspended in working oil for titanium carbide deposition process. Int J Adv Manuf Technol 40(11–12):1093–1101. https://doi.org/10.1007/s00170-008-1420-x
3. Muthuramalingam T, Mohan B (2015) A review on influence of electrical process parameters in EDM process. Arch Civ Mech Eng 15(1):87–94. https://doi.org/10.1016/J.ACME.2014.02.009
4. Rahang M, Patowari PK (2016) Parametric optimization for selective surface modification in EDM using Taguchi analysis. Mater Manuf Processes 31(4):422–431. https://doi.org/10.1080/10426914.2015.1037921
5. Ahmed N, Ishfaq K, Rafaqat M, Pervaiz S, Anwar S, Salah B (2019) EDM of Ti-6Al-4V: Electrode and polarity selection for minimum tool wear rate and overcut. Mater Manuf Processes 34(7):769–778. https://doi.org/10.1080/10426914.2019.1594278

6. Kumar SS, Varol T, Canakci A, Kumaran ST, Uthayakumar M (2021) A review on the performance of the materials by surface modification through EDM. Int J Lightweight Mater Manuf 4(1):127–144. https://doi.org/10.1016/j.ijlmm.2020.08.002
7. Bhattacharya A, Batish A, Kumar N (2013) Surface characterization and material migration during surface modification of die steels with silicon, graphite and tungsten powder in EDM process. J Mech Sci Technol 27(1):133–140. https://doi.org/10.1007/s12206-012-0883-8
8. Amorim FL, Dalcin VA, Soares P, Mendes LA (2017) Surface modification of tool steel by electrical discharge machining with molybdenum powder mixed in dielectric fluid. Int J Adv Manuf Technol 91(1):341–350. https://doi.org/10.1007/s00170-016-9678-x
9. Jabbaripour B, Sadeghi MH, Faridvand S, Shabgard MR (2012) Investigating the effects of EDM parameters on surface integrity, MRR and TWR in machining of Ti–6Al–4V. Mach Sci Technol 16(3):419–444. https://doi.org/10.1080/10910344.2012.698971
10. Bhaumik M, Maity K (2018) Effect of different tool materials during EDM performance of titanium grade 6 alloy. Eng Sci Technol Int J 21(3):507–516. https://doi.org/10.1016/j.jestch.2018.04.018
11. Samuel MP, Philip PK (1996) Properties of compacted, pre-sintered and fully sintered electrodes produced by powder metallurgy for electrical discharge machining. Ind J Eng Mater Sci 3:229–233
12. Pantelis DI, Bouyiouri E, Kouloumbi N, Vassiliou P, Koutsomichalis A (2002) Wear and corrosion resistance of laser surface hardened structural steel. Surf Coat Technol 161(2–3):125–134. https://doi.org/10.1016/S0257-8972(02)00495-4
13. Wang ZL, Fang Y, Wu PN, Zhao WS, Cheng K (2002) Surface modification process by electrical discharge machining with a Ti powder green compact electrode. J Mater Process Technol 129(1–3):139–142. https://doi.org/10.1016/S0924-0136(02)00597-6
14. Gangadhar A, Shunmugam MS, Philip PK (1991) Surface modification in electrodischarge processing with a powder compact tool electrode. Wear 143(1):45–55. https://doi.org/10.1016/0043-1648(91)90084-8
15. Mohri N, Saito N, Tsunekawa Y, Kinoshita N (1993) Metal surface modification by electrical discharge machining with composite electrode. CIRP Ann 42(1):219–222. https://doi.org/10.1016/S0007-8506(07)62429-9
16. Simao J, Lee HG, Aspinwall DK, Dewes RC, Aspinwall EM (2003) Workpiece surface modification using electrical discharge machining. Int J Mach Tools Manuf 43(2):121–128. https://doi.org/10.1016/S0890-6955(02)00187-6
17. Shunmugam MS, Philip PK, Gangadhar A (1994) Improvement of wear resistance by EDM with tungsten carbide P/M electrode. Wear 171(1–2):1–5. https://doi.org/10.1016/0043-1648(94)90340-9
18. Machkale PD, Dabade BM (2020) Morphological studies and wear behavior of electro discharge coated steel. Mater Today: Proc 27:640–648. https://doi.org/10.1016/J.MATPR.2020.01.292
19. Gill AS, Kumar S (2016) Surface roughness and microhardness evaluation for EDM with Cu–Mn powder metallurgy tool. Mater Manuf Processes 31(4):514–521. https://doi.org/10.1080/10426914.2015.1070412
20. Kruth IP, Stevens I, Froyen I, Lauwers I (1995) Study of the white layer of a surface machined by die-sinking electro-discharge machining. CIRP Ann 44(1):169–172. https://doi.org/10.1016/S0007-8506(07)62299-9
21. Ho SK, Aspinwall DK, Voice W (2007) Use of powder metallurgy (PM) compacted electrodes for electrical discharge surface alloying/modification of Ti–6Al–4V alloy. J Mater Process Technol 191(1–3):123–126. https://doi.org/10.1016/J.JMATPROTEC.2007.03.003
22. Patowari PK, Mishra UK, Saha P, Mishra PK (2010) Surface modification of C40 steel using WC–Cu P/M green compact electrodes in EDM. Int J Manuf Technol Manage 21(1–2):83–98. https://doi.org/10.1504/IJMTM.2010.034288
23. Hwang YL, Kuo CL, Hwang SF (2010) The coating of TiC layer on the surface of nickel by electric discharge coating (EDC) with a multi-layer electrode. J Mater Process Technol 210(4):642–652. https://doi.org/10.1016/J.JMATPROTEC.2009.11.013

24. Patowari PK, Saha P, Mishra PK (2010) Taguchi analysis of surface modification technique using W–Cu powder metallurgy sintered tools in EDM and characterization of the deposited layer. Int J Adv Manuf Technol 54(5–8):593–604. https://doi.org/10.1007/S00170-010-2966-Y
25. Kung KY, Horng JT, Chiang KT (2007) Material removal rate and electrode wear ratio study on the powder mixed electrical discharge machining of cobalt-bonded tungsten carbide. Int J Adv Manuf Technol 40(1, 2):95–104. https://doi.org/10.1007/S00170-007-1307-2
26. Kansal HK, Singh S, Kumar P (2007) Effect of silicon powder mixed EDM on machining rate of AISI D2 die steel. J Manuf Process 9(1):13–22. https://doi.org/10.1016/S1526-6125(07)70104-4
27. Furutania K, Saneto A, Takezawa H, Mohri N, Miyake H (2001) Accretion of titanium carbide by electrical discharge machining with powder suspended in working fluid. Precis Eng 25(2):138–144. https://doi.org/10.1016/S0141-6359(00)00068-4
28. Garg RK, Singh KK, Sachdeva A, Sharma VS, Ojha K, Singh S (2010) Review of research work in sinking EDM and WEDM on metal matrix composite materials. Int J Adv Manuf Technol 50(5):611–624. https://doi.org/10.1007/S00170-010-2534-5
29. Zhang Y, Liu Y, Shen Y, Ji R, Cai B, Li H, Wang F (2012) A review of the current understanding and technology of powder mixed electrical discharge machining (PMEDM). In: 2012 IEEE International conference on mechatronics and automation. IEEE, pp 2240–2247. https://doi.org/10.1109/ICMA.2012.6285692
30. Peças P, Henriques E (2007) Effect of the powder concentration and dielectric flow in the surface morphology in electrical discharge machining with powder-mixed dielectric (PMD-EDM). Int J Adv Manuf Technol 37(11):1120–1132. https://doi.org/10.1007/S00170-007-1061-5
31. Sahu SK, Jadam T, Datta S, Nandi G (2018) Effect of using SiC powder-added dielectric media during electro-discharge machining of Inconel 718 superalloys. J Braz Soc Mech Sci Eng 40(7):1–19. https://doi.org/10.1007/S40430-018-1257-7/FIGURES/18
32. Kansal HK, Singh S, Kumar P (2007) Technology and research developments in powder mixed electric discharge machining (PMEDM). J Mater Process Technol 184(1–3):32–41. https://doi.org/10.1016/J.JMATPROTEC.2006.10.046
33. Mohanty S, Kumar V, Kumar Das A, Dixit AR (2019) Surface modification of Ti-alloy by micro-electrical discharge process using tungsten disulphide powder suspension. J Manuf Process 37:28–41. https://doi.org/10.1016/J.JMAPRO.2018.11.007
34. Wu KL, Yan BH, Huang FY, Chen SC (2005) Improvement of surface finish on SKD steel using electro-discharge machining with aluminum and surfactant added dielectric. Int J Mach Tools Manuf 45(10):1195–1201. https://doi.org/10.1016/J.IJMACHTOOLS.2004.12.005
35. Singh AK, Kumar S, Singh VP (2014) Effect of the addition of conductive powder in dielectric on the surface properties of superalloy Super Co 605 by EDM process. Int J Adv Manuf Technol 77(1):99–106. https://doi.org/10.1007/S00170-014-6433-Z
36. Luo YF (1997) The dependence of interspace discharge transitivity upon the gap debris in precision electrodischarge machining. J Mater Process Technol 68(2):121–131. https://doi.org/10.1016/S0924-0136(96)00019-2
37. Ojha K, Garg RK, Singh KK (2012) An investigation into the effect of nickel micro powder suspended dielectric and varying triangular shape electrodes on EDM performance measures of EN-19 steel. Int J Mechatron Manuf Syst 5(1):66–92. https://doi.org/10.1504/IJMMS.2012.046144
38. Kumar S, Batra U (2012) Surface modification of die steel materials by EDM method using tungsten powder-mixed dielectric. J Manuf Process 14(1):35–40. https://doi.org/10.1016/J.JMAPRO.2011.09.002
39. Bai CY, Koo CH (2006) Effects of kerosene or distilled water as dielectric on electrical discharge alloying of superalloy Haynes 230 with Al–Mo composite electrode. Surf Coat Technol 200(12–13):4127–4135. https://doi.org/10.1016/J.SURFCOAT.2005.03.022
40. Beri N, Maheshwari S, Sharma C, Kumar A (2008) Performance evaluation of powder metallurgy electrode in electrical discharge machining of AISI D2 steel using Taguchi method. Int J Mech Mechatron Eng 2(2)

41. Li C, Xu X, Li Y, Tong H, Ding S, Kong Q, Zhao L, Ding J (2019) Effects of dielectric fluids on surface integrity for the recast layer in high speed EDM drilling of nickel alloy. J Alloys Comps 783:95–102. https://doi.org/10.1016/J.JALLCOM.2018.12.283
42. Kunieda M, Yoshida M (1997) Electrical discharge machining in gas. CIRP Ann 46(1):143–146. https://doi.org/10.1016/S0007-8506(07)60794-X
43. Lee SH, Li XP (2001) Study of the effect of machining parameters on the machining characteristics in electrical discharge machining of tungsten carbide. J Mater Process Technol 115(3):344–358. https://doi.org/10.1016/S0924-0136(01)00992-X
44. Kumar S, Singh R, Singh TP, Sethi BL (2009) Surface modification by electrical discharge machining: a review. J Mater Process Technol 209(8):3675–3687. https://doi.org/10.1016/J.JMATPROTEC.2008.09.032
45. Beri N, Maheshwari S, Sharma C, Kumar A (2010) Technological advancement in electrical discharge machining with powder metallurgy processed electrodes: a review. Mater Manuf Processes 25(10):1186–1197. https://doi.org/10.1080/10426914.2010.512647
46. Reza M, Azhari A, Tomadi SH, Hassan MA (2010) Effects of polarity parameter on machining of tool steel workpiece using electrical discharge machining. In: National conference in mechanical engineering research and postgraduate students (1st NCMER 2010), pp 621–626
47. Tsai HC, Yan BH, Huang FY (2003) EDM performance of Cr/Cu-based composite electrodes. Int J Mach Tools Manuf 43(3):245–252. https://doi.org/10.1016/S0890-6955(02)00238-9
48. Samuel MP, Philip PK (1997) Power metallurgy tool electrodes for electrical discharge machining. Int J Mach Tools Manuf 37(11):1625–1633. https://doi.org/10.1016/S0890-6955(97)00006-0
49. Gill AS, Kumar S (2015) Surface alloying by powder metallurgy tool electrode using EDM process. Mater Today: Proc 2(4–5):1723–1730. https://doi.org/10.1016/J.MATPR.2015.07.006
50. Wong YS, Lim LC, Rahuman I, Tee WM (1998) Near-mirror-finish phenomenon in EDM using powder-mixed dielectric. J Mater Process Technol 79(1–3):30–40. https://doi.org/10.1016/S0924-0136(97)00450-0
51. Yan BH, Lin YC, Huang FY, Wang CH (2001) Surface modification of SKD 61 during EDM with metal powder in the dielectric. Mater Trans 42(12):2597–2604. https://doi.org/10.2320/MATERTRANS.42.2597
52. Wong YS, Lim LC, Lee LC (1995) Effects of flushing on electro-discharge machined surfaces. J Mater Process Technol 48(1–4):299–305. https://doi.org/10.1016/0924-0136(94)01662-K
53. Lonardo PM, Bruzzone AA (1999) Effect of flushing and electrode material on die sinking EDM. CIRP Ann 48(1):123–126. https://doi.org/10.1016/S0007-8506(07)63146-1
54. Erden A (1983) Role of dielectric flushing on electric discharge machining performance. In: Proceedings of the twenty-third international machine tool design and research conference, pp 283–289. https://doi.org/10.1007/978-1-349-06546-2_34
55. Li L, Gu L, Xi X, Zhao W (2011) Influence of flushing on performance of EDM with bunched electrode. Int J Adv Manuf Technol 58(1):187–194. https://doi.org/10.1007/S00170-011-3357-8
56. Masuzawa T, Cui X, Taniguchi N (1992) Improved jet flushing for EDM. CIRP Ann 41(1):239–242. https://doi.org/10.1016/S0007-8506(07)61194-9
57. Joshi S, Govindan P, Malshe A, Rajurkar K (2011) Experimental characterization of dry EDM performed in a pulsating magnetic field. CIRP Ann 60(1):239–242. https://doi.org/10.1016/J.CIRP.2011.03.114
58. Ho KH, Newman ST (2003) State of the art electrical discharge machining (EDM). Int J Mach Tools Manuf 43(13):1287–1300. https://doi.org/10.1016/S0890-6955(03)00162-7
59. Prihandana GS, Mahardika M, Hamdi M, Wong YS, Mitsui K (2009) Effect of micro-powder suspension and ultrasonic vibration of dielectric fluid in micro-EDM processes—Taguchi approach. Int J Mach Tools Manuf 49(12–13):1035–1041. https://doi.org/10.1016/J.IJMACHTOOLS.2009.06.014
60. Ghoreishi M, Atkinson J (2002) A comparative experimental study of machining characteristics in vibratory, rotary and vibro-rotary electro-discharge machining. J Mater Process Technol 120(1–3):374–384. https://doi.org/10.1016/S0924-0136(01)01160-8

61. Zhang QH, Zhang JH, Deng JX, Qin Y, Niu ZW (2002) Ultrasonic vibration electrical discharge machining in gas. J Mater Process Technol 129(1–3):135–138. https://doi.org/10.1016/S0924-0136(02)00596-4
62. Prihandana GS, Mahardika M, Hamdi M, Mitsui K (2011) Effect of low-frequency vibration on workpiece in EDM processes. J Mech Sci Technol 25(5):1231–1234. https://doi.org/10.1007/S12206-011-0307-1
63. Xu MG, Zhang JH, Li Y, Zhang QH, Ren SF (2009) Material removal mechanisms of cemented carbides machined by ultrasonic vibration assisted EDM in gas medium. J Mater Process Technol 209(4):1742–1746. https://doi.org/10.1016/J.JMATPROTEC.2008.04.031
64. Lin CL, Lin JL, Ko TC (2002) Optimisation of the EDM process based on the orthogonal array with fuzzy logic and grey relational analysis method. Int J Adv Manuf Technol 19(4):271–277. https://doi.org/10.1007/S001700200034
65. Dewangan S, Gangopadhyay S, Biswas CK (2015) Multi-response optimization of surface integrity characteristics of EDM process using grey-fuzzy logic-based hybrid approach. Eng Sci Technol Int J 18(3):361–368. https://doi.org/10.1016/J.JESTCH.2015.01.009
66. Takagi T, Sugeno M (1985) Fuzzy identification of systems and its applications to modeling and control. IEEE Trans Syst Man Cybern SMC-15(1):116–132. https://doi.org/10.1109/TSMC.1985.6313399
67. Puertas I, Luis CJ (2004) A study of optimization of machining parameters for electrical discharge machining of boron carbide. Mater Manuf Processes 19(6):1041–1070. https://doi.org/10.1081/AMP-200035200
68. Deb K, Agrawal S, Pratap A, Meyarivan T (2000) A fast elitist non-dominated sorting genetic algorithm for multi-objective optimization: NSGA-II. Lecture Notes in Comp Sci (including subseries Lecture Notes in Artificial Intelligence and Lecture Notes in Bioinformatics) 1917:849–858. https://doi.org/10.1007/3-540-45356-3_83
69. Zhang G, Eddy Patuwo B, Hu MY (1998) Forecasting with artificial neural networks: the state of the art. Int J Forecast 14(1):35–62. https://doi.org/10.1016/S0169-2070(97)00044-7
70. Tsai KM, Wang PJ (2001) Predictions on surface finish in electrical discharge machining based upon neural network models. Int J Mach Tools Manuf 41(10):1385–1403. https://doi.org/10.1016/S0890-6955(01)00028-1

Estimation of Wire Surface Quality Index During Wire Electric Discharge Machining Using Image Processing Technique

Sanghamitra Das and Shrikrishna Nandkishor Joshi

1 Introduction

One of the main issues in the manufacturing sector is wire failure, which is caused by extensive wire erosion during WEDM. The degradation of the wire tool due to violent and random spark discharges vitiate the wire strength thus decreasing the productivity of the machining operation. Wire wear also reduces the quality of the machined products. Thus, prohibition of wire breakage and minimizing wire damages are important for effective WEDM process. An in-depth understanding of the wire erosion mechanism shall give a proper insight into the wire breakage phenomenon, thus predicting and preventing untimely wire failure.

The wire is subjected to excessive thermal loads due to successive discharges and attains a very high temperature, sometimes exceeding the material's melting point. This causes erosion of the wire material and decreases the cross-sectional area of the electrode, thus diminishing the wire strength. Literature shows that a sharp rise in temperature during the cutting process is the primary factor causing wire breakage. Researchers have developed numerical models to estimate the temperature rise in the wire electrode during machining. Dekeyser et al. [1] developed a fundamental thermal model to evaluate the effect of different process conditions on the temperature attained by the wire to predict the wire breakage phenomenon. An analytical solution is compared with a numerical finite difference method, and it was observed that the power dissipation exceeding the normal working conditions is an indicator of wire failure. Similar approaches were adopted by authors [2, 3] to predict the temperature contour in the wire and its contribution to wire failure. It was noted

S. Das · S. N. Joshi (✉)
Department of Mechanical Engineering, Indian Institute of Technology Guwahati, Guwahati, Assam 781039, India
e-mail: snj@iitg.ac.in

S. Das
e-mail: sdas93@iitg.ac.in

S. N. Joshi et al. (eds.), *Low Cost Manufacturing Technologies*,
https://doi.org/10.1007/978-981-19-8452-5_15

that a sharp rise in temperature with higher input power and reduced diameter of wire increases the probability of failure [4]. In a similar study, Banerjee et al. [5] predicted the three-dimensional transient temperature plot and material removed on the wire with the help of an explicit finite difference method. Saha et al. [6] were the first to employ a simple finite element model (FEM) to predict a critical variable for wire breakage. Transient thermal analysis for randomly located multiple discharges on the wire electrode was also reported by Banerjee and Prasad [7]. The effects of process parameters like discharge power input, pulse frequency, duty factor and wire speed were studied. The predicted values of peak wire temperature suggest the probability of wire rupture risk. The effect of moving wire during the WEDM operation was incorporated using moving heat source characteristics in a numerical model, which was developed for a single pulse [8]. The average temperature rise in the wire electrode was also recorded experimentally which was further used to evaluate the convective heat transfer coefficient during machining [9]. The temperature attained by the wire is affected by several factors, such as energy density, kerf geometry and dielectric pressure. Similarly, an attempt was made to measure the temperature rise in the wire electrode by using a two-colour pyrometer with an optical fibre [10].

An extensive literature survey carried out on various aspects of WEDM showed a scarce amount of studies available on the mechanism of wire erosion during machining [11, 12]. Wire erosion cannot be completely eliminated but can be kept to a minimum extent by optimizing the machining parameters. Therefore, it is important to estimate a tolerable limit of wire wear in order to reduce the frequency of wire failures. In the present study, an attempt has been made to determine a SQI of the eroded wires in order to estimate the intensity of wire surface damages and identify a tolerable limit of wire wear to avoid wire failure. The utilization of image processing (IP) technique to evaluate a wire safety index, which shall establish the tolerance level of wire damages without wire breakage, is elaborated in the paper.

2 Materials and Methods

The equipment, materials and process parameters used are presented in Table 1. The machined wire samples were studied with the help of FESEM and EDX.

2.1 Overall Methodology of Image Processing (IP) Technique

IP is a technique for converting an image to a digital format and doing various operations to it to produce an improved image or extract important information from it. Nowadays, IP is one of the fastest growing technologies, which has extensive applications in medical industry, astronomy, industrial robotics and remote sensing by satellites. The IP technique entails the following actions: (a) importing the image using an image acquisition programme; (b) assessing the image and (c) receiving the

Table 1 List of process parameters and their levels

Set-up	WEDM set-up (MAKE JK MACHINES, MODEL EC032)		
Workpiece	Titanium grade 5 alloy (Ti–6Al–4V)		
Wire tool material	Zinc-coated brass wire (0.25 mm diameter)		
Parameters	Level 1	Level 2	Level 3
Voltage, V (V)	60	85	–
Current, I (A)	4	6	8
Pulse on-time, t_{on} (μs)	4	8	16
Pulse off-time, t_{off} (μs)	2	4	6
Wire velocity, v (m/s)	3	6	9

output. There are two methods used for IP, viz. analogue and digital IP. Analogue IP is used for the hard copies like printouts and photographs. Digital IP is used to manipulate the digital images using computers. The processing of digital images can be divided into several classes: image enhancement, image restoration, image analysis and image compression.

The fundamental building block of a digital image is the pixel. A digital greyscale image in a computer is represented by pixels matrix. The light intensity of the image at a specific location is determined by the pixel intensity. Typically, an 8-bit number is used to represent the pixel values of a greyscale image; as a result, the pixel intensity value varies from 0 to 255 (2^8–1). A fully dark (black) image is represented by the value 0, and a completely light (white) image is represented by the value 255. A digital image is represented in matrix form as

$$f(x, y) = \begin{bmatrix} f(1,1) & f(1,2) & \ldots & f(1,N) \\ f(2,1) & f(2,2) & \ldots & f(2,N) \\ \cdot & \cdot & & \cdot \\ \cdot & \cdot & & \cdot \\ \cdot & \cdot & & \cdot \\ f(M,1) & f(M,2) & \ldots & f(M,N) \end{bmatrix} \quad (1)$$

The right side of this equation represents an image in digital form.

3 Results and Discussion

3.1 Estimation of a Wire Surface Quality Index (SQI) with the Aid of Image Processing (IP) Technique

In order to address the issues of failure and wear of the wire, the current work primarily focuses on assessing the erosion and damages experienced by the wire electrode during the WEDM process. The unused and degraded surface of a wire tool is shown in Figs. 1 and 2, respectively.

On the surface of the wire tool, the image reveals serious damage in the shape of craters, pits, holes and pits. To gauge the severity of the wire surface damages and establish a tolerable limit of wire wear to prevent wire failure, an attempt was made to calculate the SQI of the eroded wires. The established methodology will be helpful in determining a connection between wire damage and workpiece surface roughness. Using the IP technique, a wire SQI was calculated from the FESEM pictures of the wire electrode. An image histogram can be defined as a greyscale value distribution illustrating the occurrence frequency of each greyscale value. It is a graphical representation of the tonal distribution that quantifies the pixels with similar tonal value in a digital image. Histograms have many uses in IP like image analysis, image brightness, adjusting contrast of an image, image equalization, thresholding and computer vision. The horizontal axis of the histogram plot depicts the tonal variations; on the other hand, the vertical axis signifies the quantity of pixels in that particular tone. The pixel intensity values range from 0 to 255 in the horizontal axis for an 8-bit greyscale image. An image is scanned, and a running count of the number of

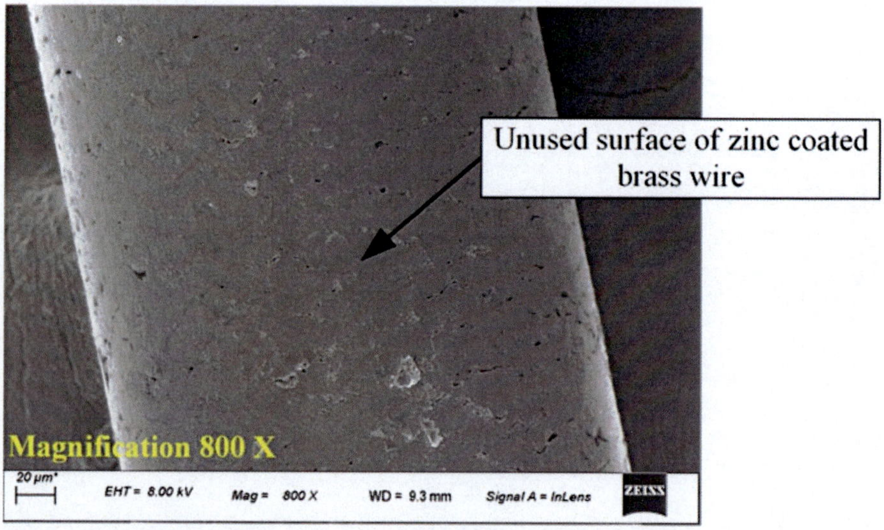

Fig. 1 FESEM image of unused zinc-coated brass wire

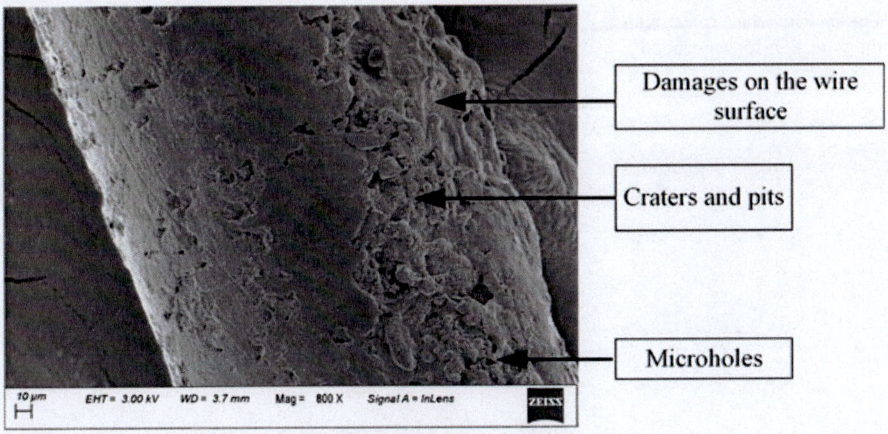

Fig. 2 FESEM image of a damaged wire surface

pixels found at each intensity value is kept. This then constructs a suitable histogram for the image. The histogram of a dark image will have most of its data points towards left and centre of the graph. On the other hand, a bright image will have its data points towards right and centre of the graph. Figure 3 depicts the flowchart for estimating the wire SQI using the IP technique. The collected wire samples were examined in FESEM. Using ImageJ software, a histogram was created for the FESEM wire pictures to detect the variation in image pixel intensity at various positions. A scale of 20 μm was kept constant while taking the FESEM images which was used while scaling the wire image in ImageJ software.

The area of the wire surface with craters or defects will indicate a different pixel intensity, making it possible to distinguish it from a surface with less wear. Thus, it can be said that an image histogram plot is an efficient tool for determining the degree erosion of the wire during the machining process. While maintaining a steady pixel count, a specific section of the wire sample was taken into consideration for examination.

The histogram mean value (HMV) of the corresponding process condition as shown in Fig. 4a was evaluated as 164.98. A similar technique was applied for Fig. 4b, which showed severe damages on the wire surfaces. The comparable histogram mean was found to be lower (118.99), indicating a shift in the histogram's axis towards the darker side. The cause of this is that the light intensity decreases in the area of the damage, which causes the pixel intensity to decrease and causes the histogram to move to the darker side. It has been noted that a wire surface image with a significant amount of damage will result in a lower mean histogram value. The histogram moves to the left (darker) side as a result of damage to the wire surface in the form of craters or pits decreasing the amount of light in that region. As a result, the HMV decreases. The level and rate of erosion of the wire can be indicated by the wire SQI, which is the mean value of the plotted histogram. However, certain amount of error could be induced in the results as the process of selecting the region in the image is manual.

Fig. 3 Flowchart to estimate the wire SQI for an eroded wire surface

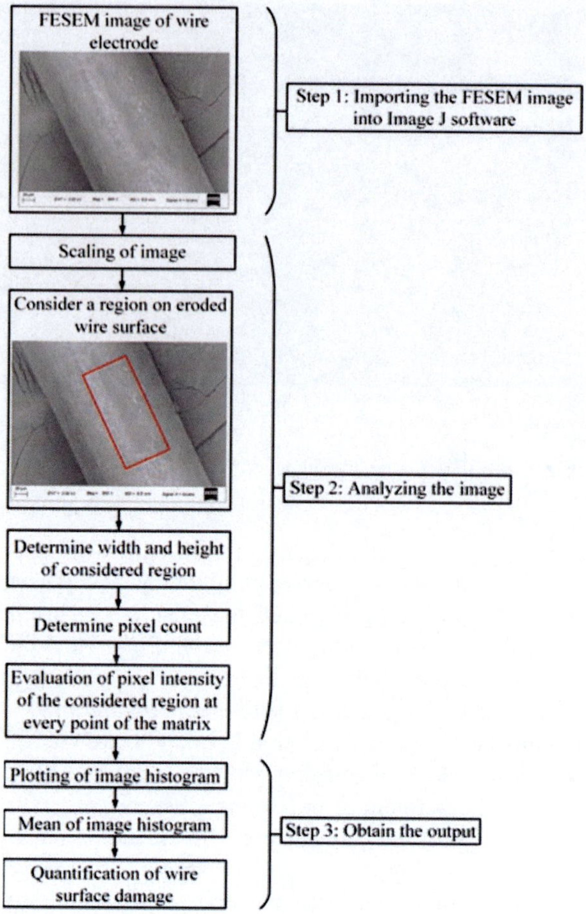

The HMVs of the FESEM wire pictures acquired under various sets of process conditions are displayed in Table 2. Greater intensity of wear on the surface of the wire is indicated by a relatively lower HMV. The established methodology can also help in comprehending the negative effects of wire erosion on product surface quality.

4 Conclusion

The conclusions that can be drawn from the current work are summarized as follows:

- An Image Processing (IP) based wire Surface Quality Index (SQI) has been proposed to evaluate the degree of erosion endured by the surface of the wire.

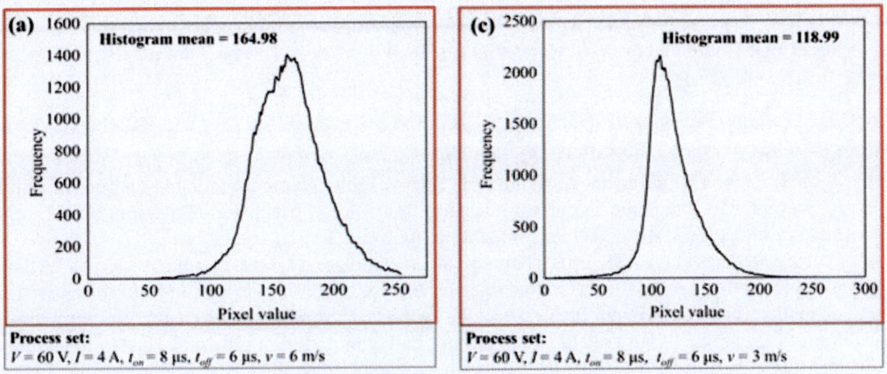

Fig. 4 Histogram plots of FESEM images of eroded wire samples at different process conditions

Table 2 HMVs of the wire samples at different machining conditions

V (V)	I (A)	t_{on} (µs)	t_{off} (µs)	v (m/s)	HMV
60	4	8	6	6	164.98
60	4	8	6	3	118.99
85	8	16	6	6	117.02
85	8	16	2	3	100.98

- The surface of the eroded zinc-coated brass wire samples was degraded in the form of craters, attachment of debris and formation of globules, micro-holes and micro-pits.
- The histogram mean value (HMV) of the eroded wire sample image was used as an indicator to find a limiting value for the wire wear, beyond which the wire is likely to break and also impairs the quality of the workpiece surface.
- Higher intensity of wire surface erosion causes a decrease in the wire SQI, i.e. lower HMVs of the wire images.

References

1. Dekeyser W, Snoeys R, Jennes M (1985) A thermal model to investigate the wire rupture phenomenon for improving performance in EDM wire cutting. J Manuf Syst 4:179–190. https://doi.org/10.1016/0278-6125(85)90024-X
2. Jennes M, Snoeys R, Dekeyser W (1984) Comparison of various approaches to model the thermal load on the EDM-wire electrode. CIRP Ann 33:93–98. https://doi.org/10.1016/S0007-8506(07)61387-0
3. Obara H, Iwata Y (1995) An attempt to detect wire temperature distribution during wire EDM. In: Proceedings of the eleventh international symposium for electro-machining (ISEM-11), Lausanne, Switzerland

4. Banerjee S, Prasad BVSSS, Mishra PK (1993) A simple model to estimate the thermal loads on an EDM wire electrode. J Mater Process Tech 39:305–317. https://doi.org/10.1016/0924-0136(93)90165-3
5. Banerjee S, Prasad BVSSS, Mishra PK (1997) Analysis of three-dimensional transient heat conduction for predicting wire erosion in the wire electrical discharge machining process. J Mater Process Technol 65:134–142. https://doi.org/10.1016/0924-0136(95)02253-8
6. Saha S, Pachon M, Ghoshal A, Schulz MJ (2004) Finite element modeling and optimization to prevent wire breakage in electro-discharge machining. Mech Res Commun 31:451–463. https://doi.org/10.1016/j.mechrescom.2003.09.006
7. Banerjee S, Prasad BVSSS (2010) Numerical evaluation of transient thermal loads on a WEDM wire electrode under spatially random multiple discharge conditions with and without clustering of sparks. Int J Adv Manuf Technol 48:571–580. https://doi.org/10.1007/s00170-009-2300-8
8. Yang X, Feng G, Teng Q (2012) Temperature field simulation of wire electrode in high-speed and medium-speed WEDM under moving heat source. Procedia CIRP 1:633–638. https://doi.org/10.1016/j.procir.2012.04.112
9. Han F, Cheng G, Feng Z, Soichiro I (2009) Measurement of wire electrode temperature in WEDM. Int J Adv Manuf Technol 41:871–879. https://doi.org/10.1007/s00170-008-1546-x
10. Koyano T, Takahashi T, Tsurutani S, Hosokawa A, Furumoto T, Hashimoto Y (2018) Temperature measurement of wire electrode in wire EDM by two-color pyrometer. In: Procedia CIRP. Elsevier B.V., pp 96–99. https://doi.org/10.1016/j.procir.2017.12.029
11. Pramanik A, Basak AK (2016) Degradation of wire electrode during electrical discharge machining of metal matrix composites. Wear 346–347:124–131. https://doi.org/10.1016/j.wear.2015.11.011
12. Pramanik A, Basak AK (2018) Sustainability in wire electrical discharge machining of titanium alloy: understanding wire rupture. J Clean Prod 198:472–479. https://doi.org/10.1016/j.jclepro.2018.07.045

Mechanical Properties of 3D Printed Modified Auxetic Structure: Experimental and Finite Element Study

Niranjan Kumar Choudhry and Biranchi Panda

1 Introduction

Cellular materials possess unique properties such as high energy absorption at low relative densities, high shear resistance, high compressive strength and tailorability makes them applicable in many engineering applications such as in automobile, medical, aerospace, robotics, sensors [1–3]. Researchers have performed quasi-static compression and studied in-plane mechanical response of these cellular structures through experimental, numerical, and theoretical approaches [4]. Further, an effort has been dedicated to improve the energy absorption performance of structures via different approaches such as the replacement of cell walls with triangular lattices [5] or combining two structures [6, 7] or adding ribs either (sinusoidal-shaped) [8] within the unit cell structure. The energy absorption can be correlated to internal cell arrangement and tailored considering auxetics geometries as an example [9].

In this study, 3D printed auxetic re-entrant honeycomb (RH) and it's modified version (proposed RH) are numerically and experimentally analyzed for their energy absorption performance. Finite element (FE)-based numerical models are used to understand the deformation behavior of the structures and results are validated by performing in-plane quasi-static compression test on 3D printed structures.

N. K. Choudhry (✉) · B. Panda
Sustainable Resources for Additive Manufacturing (SReAM) Lab, Department of Mechanical Engineering, Indian Institute of Technology Guwahati, Guwahati, Assam, India
e-mail: niran176103017@iitg.ac.in

S. N. Joshi et al. (eds.), *Low Cost Manufacturing Technologies*,
https://doi.org/10.1007/978-981-19-8452-5_16

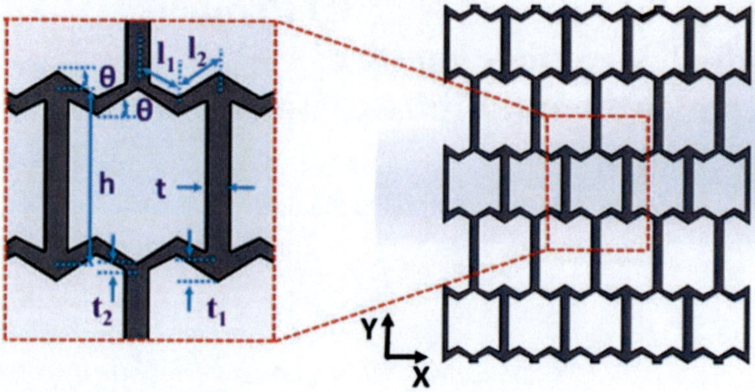

Fig. 1 Schematic of proposed RH structure with unit cell configuration

2 Materials and Methods

2.1 Structure Design

The configuration of the unit cells of the proposed RH (2D view) is depicted in Fig. 1. The proposed RH structure was designed by modifying the author's previous work [10], where inclined struts were linearly tapered (see Fig. 1). The design variables of the proposed RH structure is given in Table 1, where h is the vertical strut length, l represents the length of inclined strut, θ is angle of inclination with horizontal, and t represents the thickness of all struct in RH structure and only for vertical strut in Modified RH structure, t_1 and t_2 represent the thickness of inclined strut and l_1 and l_2 represents the length of arms. The thickness of all the samples along the Z-direction is kept constant at 35 mm.

2.2 Sample Fabrication

Figure 2 shows the proposed RH structure printed via 3D fused deposition modeling (FDM) process. The structure was first designed using Solidworks and .stl file was made for creating the tool path movement. Printing process parameters were assigned using A KISSlicer PRO v 1.5. ABS polymer filament having diameter 1.75 mm was used to extrude the material through the nozzle [11]. A nozzle with a 0.6 mm diameter was used to print the structures with a layer thickness of 0.2 mm. We set 240 °C as extrusion temperature while printing speed was at 30 mm/s with 100% infill. The "Z" direction was chosen as the building direction.

Table 1 Design variables of the unit cells of RH structures

Structure	Unit cell geometry								Specimen size			
	h (mm)	l (mm)	l_1 (mm)	l_2 (mm)	θ (°)	t (mm)	t_1 (mm)	t_2 (mm)	X (mm)	Y (mm)	Z (mm)	$\left(\frac{\rho}{\rho*}\right)$
Proposed RH	16	–	4	4	−30, 30	2	2	1	74.28	78	35	0.23

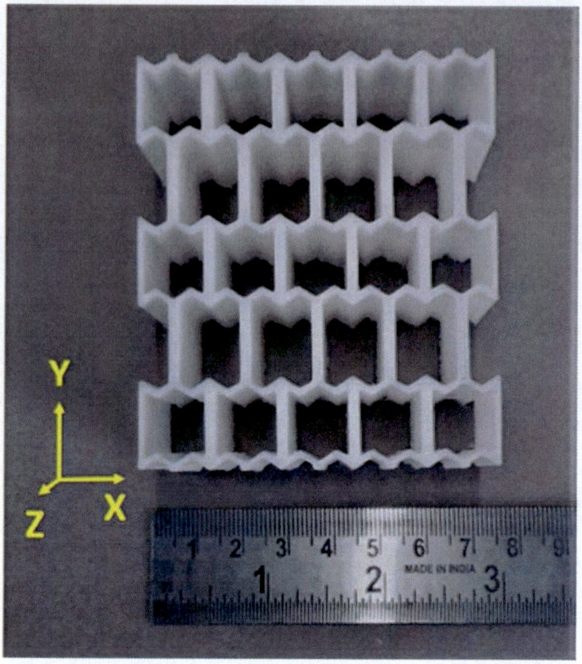

Fig. 2 FDM 3D printed proposed RH structure

2.3 Methodology

2.3.1 Experimental Procedure

The mechanical response of structures was investigated by using a quasi-static compression test at room temperature with a universal testing machine (UTM) of 250 kN capacity (MEDIAN 250). 3D printed samples were kept in-betweens the top and bottom plates. A constant displacement was given to the bottom plate at the rate of 2 mm/min (strain rate = 0.067/s) [6], while the top plate was fixed. Compression started along the *Y*-direction and was recorded using a high-definition digital camera. Load–displacement data was generated and used for further calculations. The process was repeated with three samples of each structure.

2.3.2 Numerical Simulation

The in-plane quasi-static compression of structures was simulated by using ABAQUS 2017. The numerical model was developed to mimic the experiment and get an insight into the deformation behavior of the structure and analyzed. The actual experimental conditions were replicated by putting samples in between the two rigid plates and giving loading to the *Y*-direction (Fig. 3). The material behavior of ABS was modeled by using the elastic–plastic material model. The material behavior

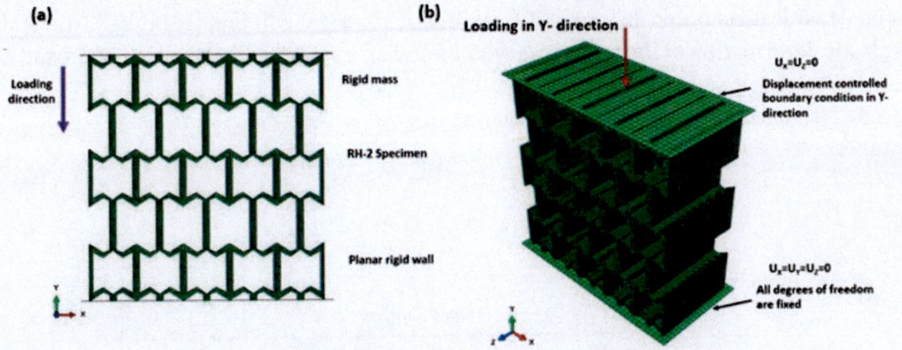

Fig. 3 Numerical modeling with boundary conditions for quasi-static compression test

Table 2 ABS polymer material properties

Matrix material	E (MPa)	σ_Y (MPa)	Poisson's ratio	ρ (g/cm^3)
ABS	2200	31	0.35	1.05

of ABS polymer is given in Table 2 and was used for material model development. "Explicit dynamic" analysis was performed with consideration of nonlinearity conditions where boundary nonlinearity, material nonlinearity and geometric nonlinearity had considered. Boundary conditions for different parts were assigned as shown in Fig. 3b. Mesh for structures was generated by using an 8-node hexahedron 3D element (C3D8R) [12] with a mesh size of 0.25 mm while a 4-node linear quadrilateral 2D element (R3D4) element of 2 mm mesh size was used to mesh rigid plates. Before the selection of mesh size mesh convergence test was performed and an optimal mesh size of 0.25 mm was chosen.

3 Results and Discussion

3.1 Numerical Model Validation

The stress–strain response of the RH and proposed RH structures are shown in Figs. 4a and 5a. It can be seen that both conventional and proposed RH structures have gone through elastic and plastic deformation till it reaches densification. The deformation behavior can be clearly distinguished into three stages [13]. At the start, a linear small region, the elastic regime was observed where stress and strains are proportional, and thereafter after reaching to the yield point, a large plateau area (where maximum energy absorption occurred by structure) was formed (Plateau regime) where structure deforms plastically. At the end of the plateau regime, the structures collapsed completely and densification regime started with the onset of the densification point. In general, the numerical model can predict both elastic and plastic deformation of the structures with a good agreement with experimental results.

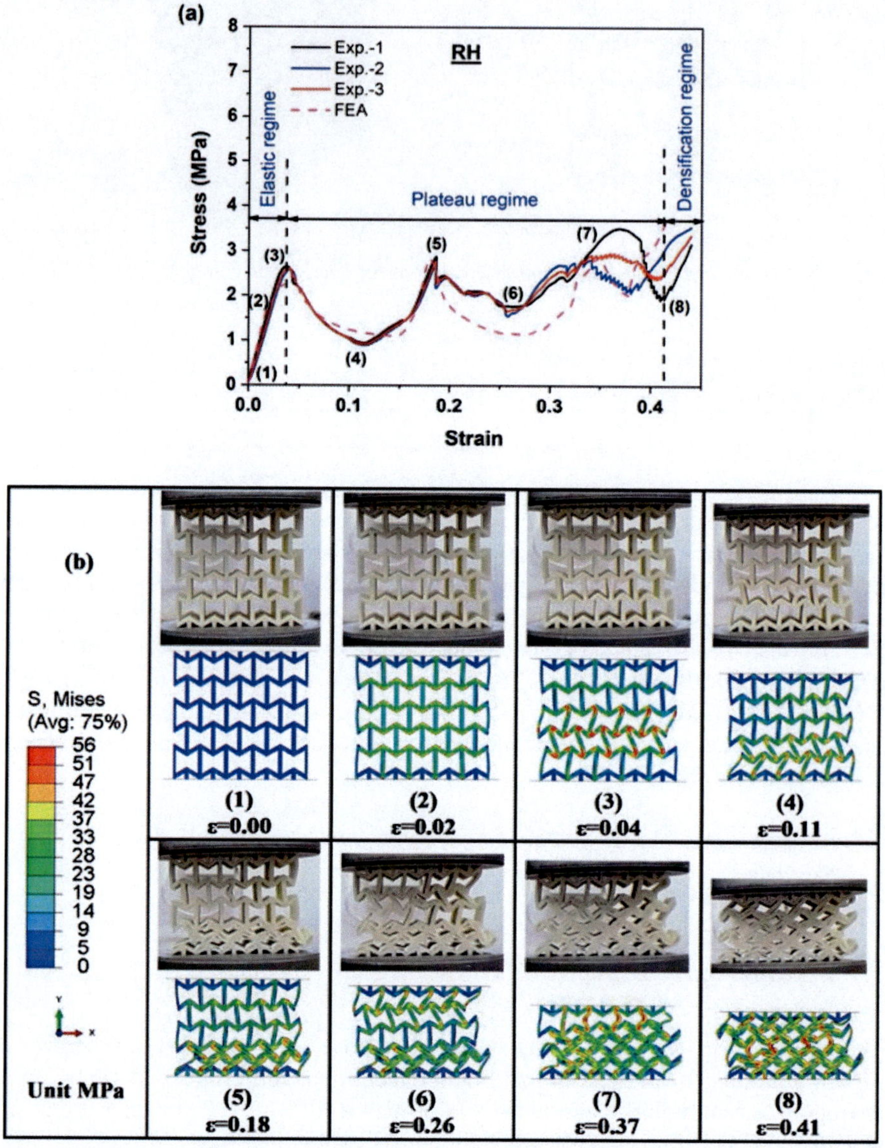

Fig. 4 RH structure **a** stress–strain response **b** experimental and FE simulation images of sequential deformation configuration (the number shown in fig. b represents the corresponding points in stress–strain plot)

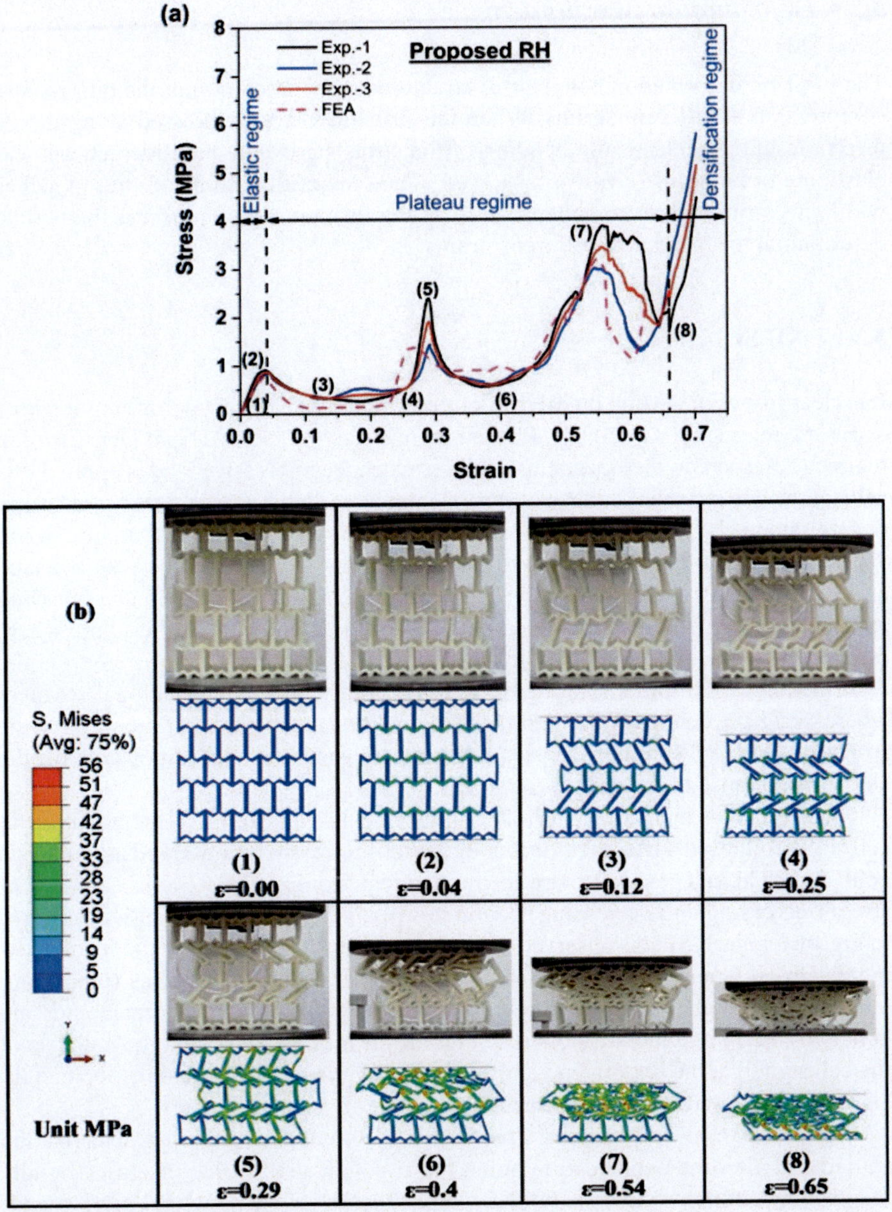

Fig. 5 Proposed RH structure **a** stress–strain response **b** experimental and FE simulation images of sequential deformation configuration (the number shown in fig. b represents the corresponding points in the stress–strain plot)

3.2 Deformation Mechanism

The in-plane deformation behavior of structures is recorded during the quasi-static compression at different strains. When the structures are compressed along the *Y*-direction, initially elastically bending of the struts starts and due to which a linear elastic deformation behavior is observed. Once the critical limit of stress reaches within the strut, it starts to collapse [13]. In this section, we will discuss the typical deformation mechanism at different strains.

3.2.1 RH Structure

It is clear from Fig. 4a that the RH structure exhibit three distinct deformation stages as Elastic regime ($0 < \varepsilon < 0.04$), Plateau regime ($0.04 < \varepsilon < 0.41$) and Densification regime ($\varepsilon > 0.41$). In the linear elastic stage, when a compressive load is applied, the cells of the RH structure deform uniformly and linearly till it reached the yield point (a maximum value). During compression at some location within the structure, when walls of the structure and loading direction became parallel then the wall behaved as an end-loaded column and buckled as long as the load crossed the Euler buckling load, while inclined cell walls bend (Fig. 4b (stage 2 and 3)). At the joints where three walls meet, the rotational stiffness played an important role to controls the deformation at the joints. The inclined walls highly affect the rotational stiffness where they have connected [13] and lead to buckling and bending of the walls of the structure. Due to the above reasons, moments are generated and responsible for the sideways collapse of the cell.

Once the stress increases to its maximum value, its value drops as the first layer of cell walls collapsed. After that, further increases in stress were observed as collapsed cells densified and resisted more to the external loading. This cyclic behavior of stress (increasing and decreasing) continues till all the rows are collapsed as shown in Fig. 4b (stage 3–5). It is observed from Fig. 4a that when the cells in the second row are collapsed (stage 6), the quasi-static engineering stress drops but its (minimum) value remains higher (than in stage 4) as more cell walls are close to each other and hence resulted in higher stiffness (see Fig. 4b). It may be noted that the collapse of the cell started at the bottom end and continuously moved forward to the top end till all layers collapsed within the structure (see Fig. 4b (stage 7 and 8)).

Finally when all cells of the structures are completely collapsed densification started and after the densification point, the resultant stress value increases rapidly due to compacting of the cells (at the compaction phase the structural performance was controlled by the base material from which the lattice structure had fabricated [12] and stiffness of lattice material are higher than the structure). Hence it can be concluded that failure of the RH structure follows a bending-dominated mechanism.

3.2.2 Proposed RH Structure

Stress–strain curve and deformation behavior of the proposed RH structure are shown in Fig. 5. Three deformation stages such as Elastic regime ($0 < \varepsilon < 0.04$), Plateau regime ($0.04 < \varepsilon < 0.65$) and Densification regime ($\varepsilon > 0.65$) are identified. The proposed RH structure has two tapered arms (l_1 and l_2) with uniformly linear tapered thicknesses (changes from 2 to 1 mm). The proposed RH structure also shows bending-dominated failure but it has a unique deformation pattern. In the elastic regime, it is observed that initially, the l_2 arm started deforming due to bending (as had lower stiffness than all other struts) and rotated about the joint of l_1 and l_2. The rotation of the strut about a node is influenced by the rotational stiffness at the joint [13]. The resistance of the structure in the early stage was low and hence has low peak stress (see Fig. 5a-(2)).

Further, as compression continued the whole structure acts like an end-loaded column and it starts failing due to buckling, and the middle layer moved outside Fig. 5b-(2–5). This behavior is not visible in the case of RH structure. Afterward, the vertical links of the 2nd and 4th row of structure (counted from the bottom) become inclined and behave as an inclined link as in the case of RH structure. They show high resistance to bending due to higher stiffness and hence the load value increases as they start to bend (see Fig. 5b-(3–5)). After collapsing of the 2nd and 4th row, the complete structure is collapsed as observed in the case of RH structure (see Fig. 5b-(6–8)).

3.3 *Mechanical Properties*

3.3.1 Modulus and Strength

The elastic modulus (E) and compressive strength [6] of a structure determine it's load bearing capacity. The mechanical properties of both structures under quasi-static compression are shown in Table 3. These properties were measured from simulation and found to be consistent with the experimental results. It can be noted from Table 3 that the proposed RH structure has the highest compressive strength than that of RH while the RH structure has the highest elastic modulus than that of the proposed RH structure and hence it has a stronger support ability during small deformation.

3.3.2 Energy Absorption

The energy absorption (EA) performance [14] of an auxetic structure is considered up to the point of the densification strain (ε_{cd}). There are many ways to calculate the densification strain (ε_{cd}) [15], and here energy efficiency method is used for calculating it. Energy absorption efficiency (η) is calculated by using Eq. (1) [15].

Table 3 Mechanical properties of structures during In-plane quasi-static compression

Structure		E (MPa)	σ_Y (MPa)	Plateau strength (MPa)
RH	Exp.-1	88.38	2.567	3.51
	Exp.-2	68.77	2.54	2.78
	Exp.-3	78.58	2.56	2.91
	FEA	78.94	2.48	2.75
Proposed RH	Exp.-1	30.01	0.91	3.85
	Exp.-2	28.45	0.83	2.98
	Exp.-3	26.93	0.87	3.36
	FEA	23.27	0.81	3.17

$$\eta = \frac{1}{\sigma_m} \int_0^{\varepsilon_m} \sigma(\varepsilon).\mathrm{d}\varepsilon \tag{1}$$

where ε_m represents strain value and σ_m represents corresponding stress value.

A curve is plotted between energy absorption efficiency and stress–strain. The corresponding crushing strain point where energy absorption efficiency is maximum is the onset of densification, (see Fig. 6). Once we get the onset of densification, Energy absorption (MJ/m^3) is calculated by using Eq. (2) [14].

$$\mathrm{EA} = \int_0^{\varepsilon_D} \sigma(\varepsilon)\mathrm{d}\varepsilon \tag{2}$$

where ε_D is the densification strain.

Total energy absorption capacity (MJ/m^3) by a structure includes the total area within the stress–strain plot starting from the beginning up to the densification strain as shown by the shaded area in Fig. 6.

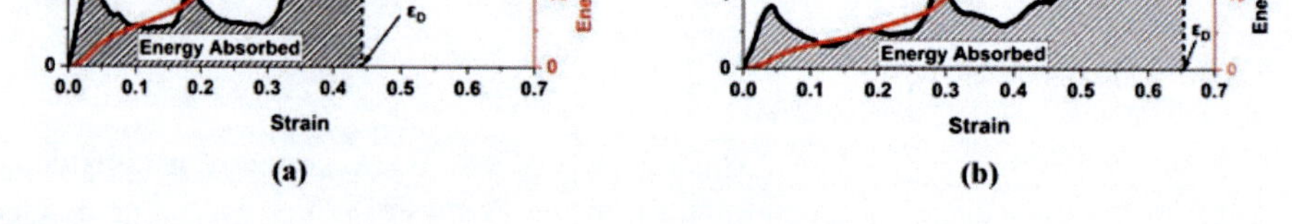

Fig. 6 Stress–strain response and energy absorption efficiency curves of **a** RH and **b** proposed RH structure

Table 4 Comparative analysis of the performance of RH structures

Structure	E (MPa)	σ_Y (MPa)	ε_{cd}	Energy absorbed (MJ/m^3)	SEA (J/kg)
RH	86.38	2.56	0.447	2.161	2084.51
Modified RH (author's previous work)	87.32	2.83	0.576	2.984	2816.47
Proposed RH	30.01	0.91	0.657	3.672	3486.32

The deformation and failure characteristics of a structure play a very crucial role in absorbing the energy before failure. It is clear from Table 4 that the proposed RH structure has absorbed (~3.672 MJ/m^3) ~65.36% more energy than that of the RH structure (~2.161 MJ/m^3) and ~23.06% more energy than modified RH structure (~2.984 MJ/m^3). While the SEA of the proposed RH structure (~3486.32 J/kg) ~67.25% more than the RH structure (~2084.51 J/kg) and ~23.79% more than the Modified RH structure (~2816.47 J/kg). The improvement in energy absorption of the proposed structure is mainly due to the improvement in the densification strain (46.98% and 14.06% improvement than that of RH and Modified RH structure, respectively).

4 Conclusion

In this paper, a comparative analysis of mechanical properties is carried out between 3D printed auxetic re-entrant structures. The structures differ from each other in terms of inclined strut thickness and the number of inclined struts. The results emerging from this study are as follows:

1. The RH and proposed RH structures exhibit similar deformation behavior but due to the unique deformation mechanism of the proposed RH, it resulted in improved energy absorption properties. It can be concluded that the deformation behavior of structures can be controlled and tailored by tailoring the design parameters.
2. The RH structure showed the highest elastic modulus, which suggests it can strongly support small deformation.
3. Energy absorption efficiency (η) of the proposed RH structure is 34.41% higher than the RH structure.
4. The energy absorption (EA) capacity and SEA of the proposed RH structure are 65.36% and 67.25% higher than the RH structure, respectively.

Overall, the findings revealed in this paper have provided new insights on the deformation mechanism of proposed re-entrant honeycomb auxetic structure, which can be very useful for designing of defense and protective engineering applications.

Acknowledgements The present work was funded by the Department of Science and Technology, India (No. SRG/2021/000052).

References

1. Surjadi JU, Gao L, Du H, Li X, Xiong X, Fang NX et al (2019) Mechanical metamaterials and their engineering applications. Adv Eng Mater 21:1800864
2. Yu X, Zhou J, Liang H, Jiang Z, Wu L (2018) Mechanical metamaterials associated with stiffness, rigidity and compressibility: a brief review. Prog Mater Sci 94:114–173
3. Nguyen-Van V, Choudhry NK, Panda B, Nguyen-Xuan H, Tran P (2022) Performance of concrete beam reinforced with 3D printed bioinspired primitive scaffold subjected to three-point bending. Autom Constr 134:104060
4. Ren X, Das R, Tran P, Ngo TD, Xie YM (2018) Auxetic metamaterials and structures: a review. Smart Mater Struct 27:23001
5. Lv W, Li D, Dong L (2020) Study on mechanical properties of a hierarchical octet-truss structure. Compos Struct 249:112640. https://doi.org/10.1016/j.compstruct.2020.112640
6. Ingrole A, Hao A, Liang R (2017) Design and modeling of auxetic and hybrid honeycomb structures for in-plane property enhancement. Mater Des 117:72–83. https://doi.org/10.1016/j.matdes.2016.12.067
7. Alomarah A, Masood SH, Sbarski I, Faisal B, Gao Z, Ruan D (2020) Compressive properties of 3D printed auxetic structures: experimental and numerical studies. Virtual Phys Prototyp 15:1–21. https://doi.org/10.1080/17452759.2019.1644184
8. Li D, Yin J, Dong L, Lakes RS (2018) Strong re-entrant cellular structures with negative Poisson's ratio. J Mater Sci 53:3493–3499. https://doi.org/10.1007/s10853-017-1809-8
9. Zhang X, Ding H, An L, Wang X (2015) Numerical investigation on dynamic crushing behavior of auxetic honeycombs with various cell-wall angles. Adv Mech Eng 7:679678
10. Choudhry NK, Panda B, Kumar S (2022) In-plane energy absorption characteristics of a modified re-entrant auxetic structure fabricated via 3D printing. Compos Part B Eng 228:109437
11. Choudhry NK, Bankar SR, Panda B, Singh H (2022) Experimental and numerical analysis of the bending behavior of 3D printed modified auxetic sandwich structures. Mater Today Proc 56:1356–1363
12. Bai L, Gong C, Chen X, Sun Y, Xin L, Pu H et al (2020) Mechanical properties and energy absorption capabilities of functionally graded lattice structures: experiments and simulations. Int J Mech Sci 182. https://doi.org/10.1016/j.ijmecsci.2020.105735
13. Gibson LJ, Ashby MF (1999) Cellular solids: structure and properties. Cambridge University Press
14. Shokri Rad M, Hatami H, Alipouri R, Farokhi Nejad A, Omidinasab F (2019) Determination of energy absorption in different cellular auxetic structures. Mech Ind 20:15–20. https://doi.org/10.1051/meca/2019019
15. Li QM, Magkiriadis I, Harrigan JJ (2006) Compressive strain at the onset of densification of cellular solids. J Cell Plast 42:371–392

Response of Coconut Coir Filler-Reinforced Epoxy Composite Toward Cyclic Loading: Fatigue Property Evaluation

Faladrum Sharma, Rahul Kumar, and Sumit Bhowmik

1 Introduction

Recently, composites reinforced with natural fibers such as coir, bamboo, hemp, flax, and ramie have received significant importance both in terms of industrial applications and research owing to ecological advantages. Public attention is now shifted on environmental gentle composites as compared to synthetic fibers. The increasing environmental and sustainability concern has raised awareness in using natural fibers as reinforcement in polymer composites instead of synthetic fibers such as glass and carbon [1, 2]. Some benefits of natural fibers comprise low price and density, ease of availability, sustainability as well as biodegradability and negligible abrasive wear of processing equipment. Also, the natural fibers maintain the ecological equilibrium due to the balanced CO_2 emission and acceptable energy recovery [3, 4].

Despite potential remunerations, they possess a certain deficit in mechanical properties and resistance to weathering. Cellulose, hemicelluloses, lignin, and other waxy substances present in their structures allow moisture absorption from the atmosphere resulting in weak bonding with the matrix [5–10]. Moreover, the differences between the chemical configurations of the fibers and matrix make the coupling between them more challenging. This results in poor stress transfer through the interface of the

F. Sharma (✉)
Department of Mechanical Engineering, Indian Institute of Technology Guwahati, Guwahati, Assam, India
e-mail: fds80@iitg.ac.in

F. Sharma · R. Kumar · S. Bhowmik
Department of Mechanical Engineering, National Institute of Technology Silchar, Silchar, Assam, India

R. Kumar
Department of Mechanical Engineering, Dayananda Sagar University, Bengaluru, Karnataka, India

S. N. Joshi et al. (eds.), *Low Cost Manufacturing Technologies*,
https://doi.org/10.1007/978-981-19-8452-5_17

composites. Hence, certain types of treatments based on the use of reagent functional groups are needed that are capable of reacting with the fibers to change their composition. This results in the reduction of moisture content and facilitates better compatibility with the matrix [6, 11, 12]. Treatment of the fibers by sodium hydroxide (NaOH) is one of the most widely used techniques to modify the molecular structure. Its advantages include ease of availability, simple processability, and safety. By providing easy access to penetrate chemicals, the hydroxide groups are changed to alkoxide groups, thereby increasing the fibers' moisture resistance property. It takes a certain portion of waxy substances including lignin and pectin [13–15]. It affects different mechanical properties like tensile, flexural, fracture, and fatigue characteristics. Among these, the fatigue properties are one of the most important as it defines the service life of the manufactured components.

The types of stresses encountered throughout the service conditions are often fluctuating and cyclic in nature. This causes fatigue degradation and hence results in the failure of the composites. Proper assessment of their fatigue behavior is essential to predict their behavior under several kinds of repetitive loading. To assess the fatigue life of polymers, Wohler (*S–N*) curves are generally used. This gives information about the number of cycles at which the material fails under different loading conditions and allows to predict the fatigue limit. The input parameters used in conducting fatigue tests are frequency, stress ratio, and loading levels based on the ultimate tensile strength (UTS) of the material. Fatigue behavior of composites reinforced with synthetic fibers like carbon, glass, and kevlar have been investigated by numerous researchers [16–20]. Those researchers conducted experiments where the specimens were subjected to several types of cyclic loading. Also, other mechanical tests, viz. tensile, bending, impact, and flexural tests were conducted. Based on the experimental results, it was found that matrix cracking, interfacial debonding, fiber pull-out, and breakage were the failure mechanisms under fatigue. However, the fatigue behavior of natural fiber composites is less investigated. In one of the investigations, Gassan [21] reported on the fatigue behavior of natural fiber-reinforced thermoset and thermoplastic matrix composite. Fatigue tests were conducted in tension–tension mode. It was found that a stronger interface results in high dynamic modulus. Also, a reduction in stiffness was observed with an increase in the number of cycles. As compared to the natural fiber, research on fatigue investigation on microparticle filler reinforcement in thermoset matrix composite material is relatively less.

The present study investigated the fatigue behavior of coir filler-reinforced epoxy composites with four different coir filler wt% along with neat epoxy samples. Four different weight percentages of untreated and treated fillers, viz. 2.5, 5, 7.5, and 10% were selected for sample preparation along with neat polymer samples. Initially, the tensile test was conducted for all filler loading samples to evaluate the ultimate tensile strength of developed material. After that, both untreated and alkali treated samples underwent tension–tension fatigue tests with three loading levels, viz. 30, 45, and 60% of ultimate tensile strength (UTS). The fatigue tests of the samples were conducted up to 10^6 cycles. In some cases, the samples underwent failure before attaining 10^6 cycles. Also, the void contents of the prepared composite samples were experimentally evaluated to assess its structural applicability.

2 Materials and Experimental Details

The methodology of this research work is described in the following subsections. The selection of proper filler material along with its extraction is described. Next, proper chemical treatment technique is discussed in details. The selection of matrix and the procedure of sample preparation are mentioned in the latter subsection. Finally, the testing procedures are discussed.

2.1 *Materials*

The coir filler (*Cocos nucifera*) of particle size less than '150 μm' was prepared from coconut husks. The coconut husk was dried and ball milled for 20 min to form filler particles. Consequently, multiple sieving was carried out to retain particle below the aforesaid size. The filler was then washed several times with distilled water. It was also washed with acetone to eliminate any foreign particles. Afterward, the washed filler was properly dried in a hot air oven at 60 °C for 8 h. Then the dried coir was packaged in an airtight polyethylene bags until needed. The coir filler has a density of 1.24 g/cm^3. In this work, the commercial grade standard epoxy adhesive 'Araldite' was chosen as the matrix material belonging to 'epoxy family' of thermosetting polymer division. Araldite comprising epoxy resin (AW 106) and hardener (HV 953U) was provided in the ratio of 10:8 by weight. As per the specification of the supplier, the specific gravity of resin is 1.17 and that of the hardener is 0.92. The viscosity of resin and hardener from the material data sheet at 25 °C are 50 Pa·s and 35 Pa·s, respectively. The high viscosity of epoxy adhesive is the basis for its selection of matrix material. It prevents micro-size fillers from settling at the bottom part of the container before curing and hence ensures uniform distribution in the matrix.

2.2 *Alkali Treatment*

The dried coir filler underwent a chemical treatment process. For this, a 5 wt% aqueous sodium hydroxide (NaOH) solution was prepared. The coir particle fillers were soaked in 5 wt% aqueous NaOH solution. Then, a beaker containing entire solution were placed on a magnetic stirrer. The solution was stirred for 8 h at 870 rpm and 48 °C. Subsequently, the fillers were washed with distilled water until a pH of 7 was attained. Those washed fillers were then dried in hot air oven for 12 h. Then it was packed in airtight polythene bags containing Silica Gel to protect the fillers

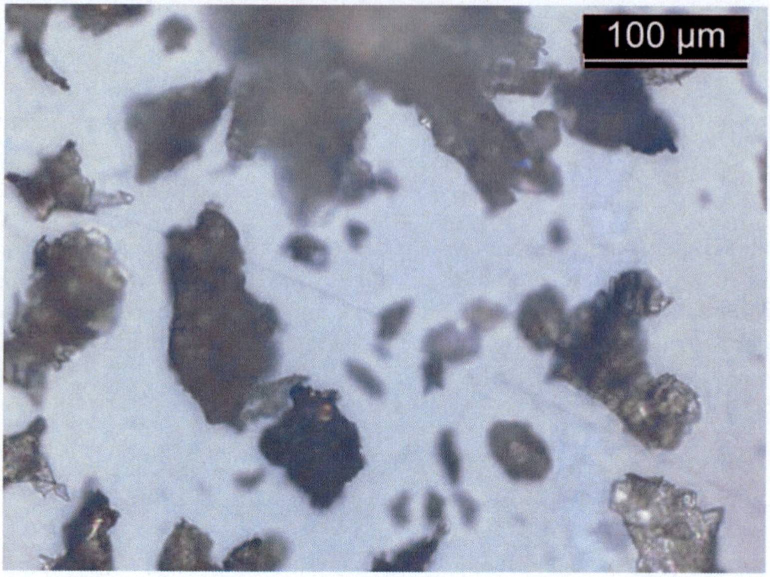

Fig. 1 Coir filler size distribution in optical micrograph

from moisture. The treated filler size distribution is schematically shown in Fig. 1 as optical micrographs. The probable chemical reaction occurred between molecules of fillers and sodium hydroxide is demonstrated as

$$\text{Filler cell}{-}\text{OH} + \text{NaOH} \rightarrow \text{filler}{-}\text{O}{-}\text{Na}^{+} + \text{H}_2\text{O} \quad (1)$$

2.3 *Sample Preparation*

The composite samples, both with untreated and treated filler reinforcements, were fabricated by a popular method referred to as hand layup technique. Four different weight percentages of untreated and treated fillers, viz. 2.5, 5, 7.5, and 10%, were selected for sample preparation along with neat polymer samples. The epoxy resin (Araldite AW 106) and the hardener (HV 953IN) were mixed in a ratio of 10:8 by weight and subsequently stirred for 3–4 min. In the next step, fillers were added to the prepared mixture. The mixture was mechanically stirred for 8–10 min in a high speed mixer at 150 rpm. Then the entire mixture was gradually poured into the glass mold of necessary dimension prepared as per ASTM standard. It was coated beforehand with overhead projector (OHP) sheet and sprayed with a leasing agent. The entire mixture in the mold was cured for 24 h at a room temperature. After this time interval of 24 h, the composite was cured. It was then properly removed from the mold.

2.4 *Procedure*

The loss in weight of the fillers before and after chemical treatment was measured. Subsequently, the void content test was performed on the samples. The mechanical tests were performed; first the tensile tests, followed by a series of fatigue tests by proper ASTM guidelines.

2.4.1 Weight Loss of Fillers

The fillers before and after treatment were tested for estimating the change in weight. The change in weight occurred due to the partial elimination of non-cellulosic components namely lignin, hemicellulose, and other waxy substances from filler surfaces owing to alkali treatment. The percentage loss was estimated by weighing a fixed amount of untreated fillers. The standard deviation was also obtained from the three identical measurements to assess the extent of repeatability. The percentage loss in the weight is evaluated as

$$\% \text{ Weight loss} = \frac{W_1 - W_2}{W_1} \times 100, \tag{2}$$

where W_1 is the initial weight of fillers and W_2 is the final weight of fillers.

2.4.2 Void Content Test

The void content test for untreated as well as treated coir filler-reinforced composite samples was performed as per ASTM D 2734-94. It is measured by determining the theoretical and actual density. Actual density is measured by specific gravity test. On the other hand, the theoretical density is evaluated as

$$D_{\mathrm{T}} = \frac{100}{\frac{R}{D} + \frac{r}{d}}, \tag{3}$$

where D_{T} is the theoretical density, R is the resin in composite (wt%), D is the density of resin, r is the reinforcement in composite (wt%) and d is the density of reinforcement. Consequently, void content is determined by

$$V = \frac{D_{\mathrm{T}} - D_{\mathrm{E}}}{D_{\mathrm{T}}} \times 100, \tag{4}$$

where V is the void content (volume %), D_{T} is the theoretical density and D_{E} is the experimental density.

Table 1 Input parameters for fatigue tests

Frequency (f)	Stress ratio (R)	Time period (t)	Amplitude ratio (A)	Loading levels (N)
5 Hz	0.5	0.2 s	0.33	1105.88 1658.82 2211.76

2.4.3 Mechanical Properties Test

The samples underwent tensile tests to determine the values of stresses for fatigue properties. The testing method for determining tensile properties of random discontinuous fiber composites was carried out as per ASTM D 3039M-14 designation. In this work, the testing was performed on a universal testing machine (Instron 8801) at a head speed of 2 mm/min. The size of the samples was 250 mm × 25 mm × 5 mm. Three specimens from different batches were tested for each filler wt%, and the mean (average) value was taken as the ultimate tensile strength (UTS) of that composite.

The tension–tension fatigue tests were performed as per ASTM D 3479M-96. The dimension of the specimen was same as that of the tensile test specimens. Experiments were conducted under the displacement-control mode at a room temperature of 22 °C and relative humidity of 56%. The different input parameters for fatigue test are presented in Table 1. The testing frequency (f) was set at a constant value of 5 Hz to avoid self-heating. Three different stress levels (30, 45, and 60% of the UTS of the composite) were applied. At each stress level, three specimens were tested. During the experiments for fatigue testing, the specimens underwent numerous number of cycles. The upper grip of the universal testing machine was movable, while the lower grip was fixed throughout the testing. For each sample, the number of cycles to failure was recorded in the computer with data acquisition system. As mentioned in Sect. 1, tests were carried up to 10^6 cycles or a particular specimen's failure, whatever comes first.

3 Results and Discussion

The justifications of the tests are outlined in the following subsections. The reasons for the weight loss of the alkali treated fillers, variation in void for different filler content, are explained. Lastly, the effects and consequences of tensile and fatigue tests due to various parameters are discussed.

Table 2 Loss in weight of the alkali treated coir fillers

Initial weight (W_1) (g)	Final weight (W_2) (g)	Weight difference (W_1–W_2) (g)	Weight loss (%)	Std. dev.
20.20	19.89	0.31	1.53	0.045

3.1 *Weight Loss of Coir Filler*

The weight of the fillers was taken before and after the chemical treatment to quantify the effect of alkali treatment [22, 23]. The weight loss of the fillers was evaluated after 8 h of alkali treatment. The values along with standard deviation are given in Table 2. The loss in weight of the alkali treated fillers might be due to dissolution of the hemicellulose and lignin content. Some amount of wax layer along the filler surface was also washed away. The aqueous NaOH solution alters the orientation of the packed crystalline cellulose order and forms an amorphous region. This provides an easy access to penetrate chemicals. Micro-molecules of the cellulose are separated at large distances in the amorphous region. Also, the alkali sensitive hydroxyl molecules (OH) breaks down. These hydroxyl molecules react with water molecules (H–OH) and eventually move out from filler structures. The remaining reactive molecules form filler cell–O–Na groups between the cellulose molecular chains. The hydroxide functional group gets converted into the alkoxide group.

3.2 *Void Content*

A difference in the values between the experimental and the theoretical density is observed. This is due to the presence of voids and pores in the prepared composite samples. Voids are principally present due to the presence of moisture in the atmosphere in the course of manufacturing as well as air trapped during the mixing. However, in some cases despite the degassing, voids are unavoidable. The measured values of void contents for untreated and treated filler composite samples are demonstrated in Fig. 2. Here, the error bar denotes the standard deviation in each case. With an increase of filler content, void content also increased. It is mainly due to the clustering of the fillers for which the resin and the hardener could not mix properly. Another observation can be built up regarding lower void contents in the case of treated filler composite specimens. This is due to the reduced hydrophilic property of coir filler owing to alkali treatment. It signifies reduced moisture absorption and porosity. The maximum void content of the prepared composite samples was found to be 2.94% for 10 wt% of untreated filler. On the other hand, the minimum value of the void content was 1.59% for 2.5% of treated filler content. These voids have enormous effect on the performance of composites. Higher void content results in poor mechanical properties including lower fatigue resistance.

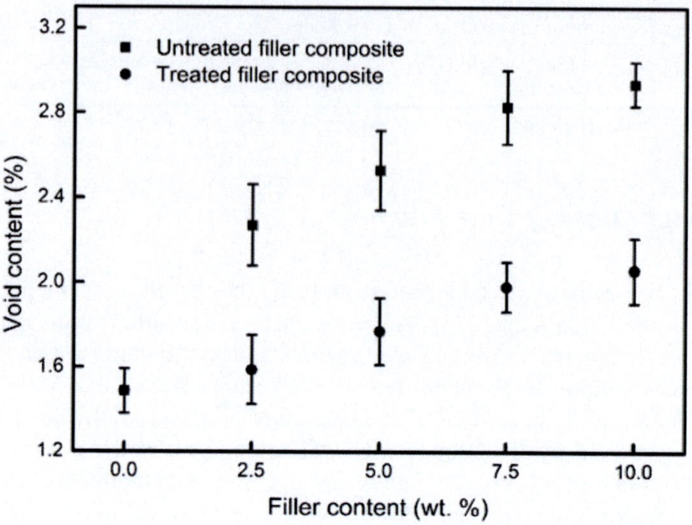

Fig. 2 Void content with the dependence of filler content for coir filler-reinforced epoxy composites

3.3 *Tensile Properties*

The values of UTS for the untreated, as well as treated filler composite samples, are demonstrated in Figs. 3 and 4, respectively. The error bars depict the standard deviation. It was found that with an increase of filler content (wt%), the tensile strength of the composite samples increased. The enhancement in tensile strength is due to the easy and effective transfer of load from filler to matrix in the presence of coir fillers [15, 24]. However, after a certain amount of filler addition in the epoxy matrix, a decline of tensile strength was observed. This is due to the poor stress transfer between the fillers and the matrix owing to the increased filler density. It results in filler agglomeration and nominal load transfer as also observed by other researchers [25]. The minimum value of tensile strength was 19 MPa for 0 wt% of filler content, i.e., for pure epoxy samples. On the other hand, the maximum tensile strength was 22 MPa for 10 wt% of coir filler. Similar observations were found in the work of other researchers [26, 27].

For treated filler reinforcement as shown in Fig. 4, the tensile strength of the composite samples increased due to NaOH treatment. An enhancement in the tensile strength from 19.3 MPa to 20.1 MPa was observed corresponding to 2.5 wt% of filler. The UTS for 5 wt% of filler is 24 MPa. The increase in the tensile strength is due to the removal of hemicellulose. Also, some portion of lignin removed after alkali treatment increases the interfacial adhesion between the matrix and the treated filler resulting in effective stress transfer. However, the increase in strength from 5 wt% of filler to 7.5 wt% of filler is not significant as compared to earlier cases. This is due to the increased brittleness of the fillers after NaOH treatment, the fillers

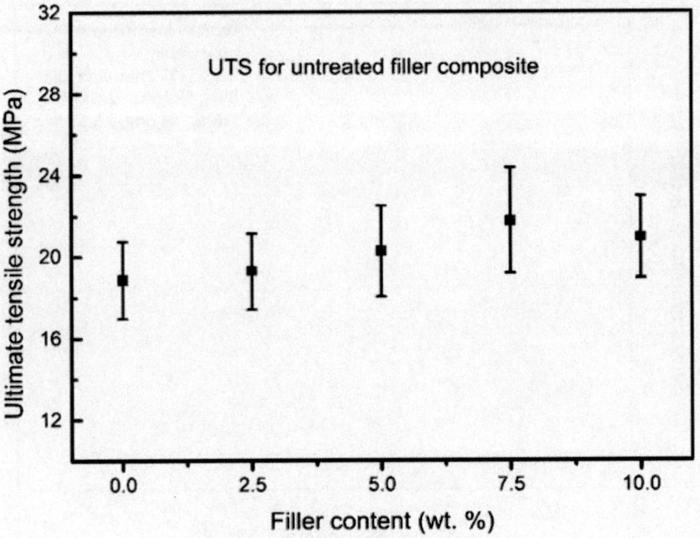

Fig. 3 Variation of UTS with the dependence of filler content for untreated filler composite

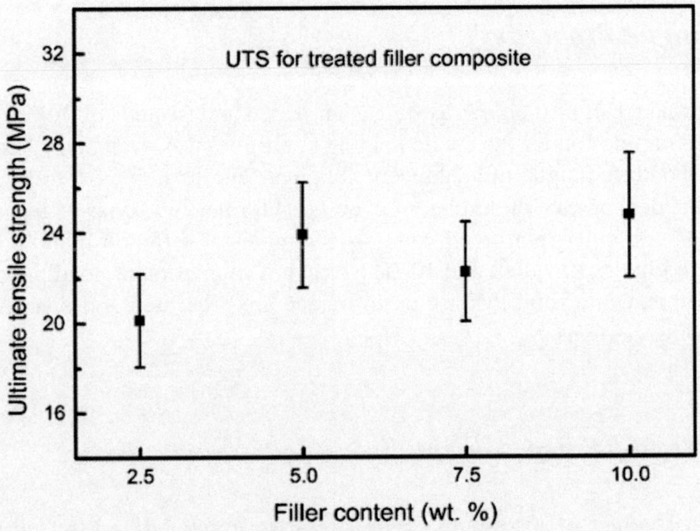

Fig. 4 Variation of UTS with the dependence of filler content for treated filler composite

suffered breakage resulting in deficient stress transfers. The maximum value of UTS is 25 MPa for 10 wt% of coir treated filler.

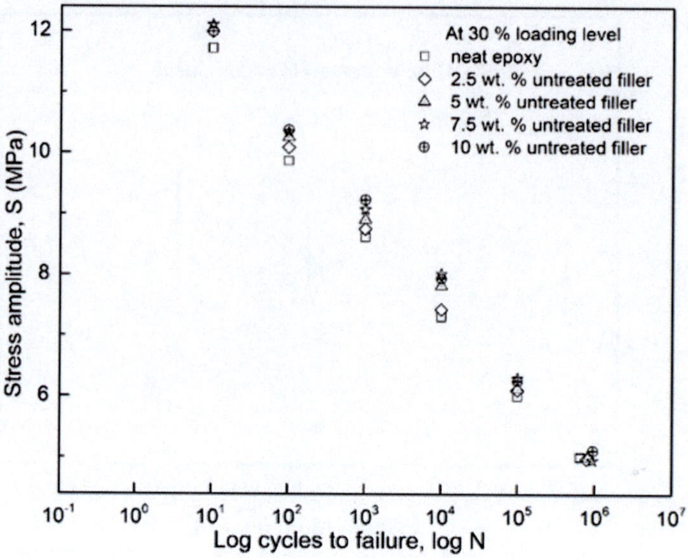

Fig. 5 *S–N* curve at 30% loading for untreated filler composite

3.4 *Fatigue Properties*

The performance of neat epoxy, as well as untreated and treated coir filler-reinforced composite samples under the cyclic loading condition, was evaluated. Accordingly, the corresponding fatigue life data were obtained. The variation of stress amplitude (S) with the dependence of number of cycles (N) for neat epoxy as well as untreated and treated coir filler-reinforced composite samples at different loading levels are depicted in Figs. 5, 6, 7, 8, 9 and 10. The effect of filler content, loading levels, and alkali treatment on cyclic loading performance are presented and discussed under the following captions.

3.5 *Effect of Filler Content*

The distinct values of maximum cyclic stress for untreated and treated coir filler composite material at 30%, 45%, and 60% of loading conditions are provided in Tables 3 and 4, respectively. The consequence of coir filler inclusion in the epoxy matrix resulted in an increase in the values of cyclic stress for all loading conditions. The maximum value of cyclic stress for neat epoxy samples at 30% loading level is 11.72 MPa. At the same loading level, the value of stress for composite samples of 2.5% untreated filler content is 11.991 MPa. The magnitude of stresses for other untreated filler content like 5%, 7.5%, and 10% at the same loading level are 12.042 MPa, 12.104 MPa, and 11.992 MPa, respectively. With the increase of filler

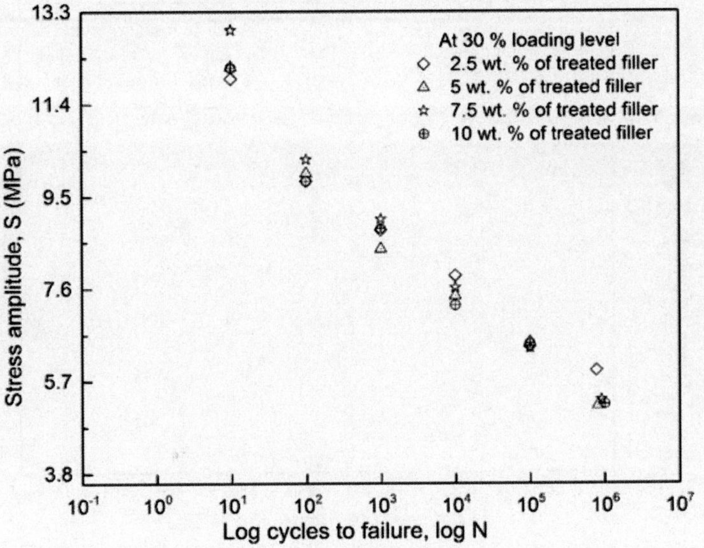

Fig. 6 *S–N* curve at 30% loading for treated filler composite

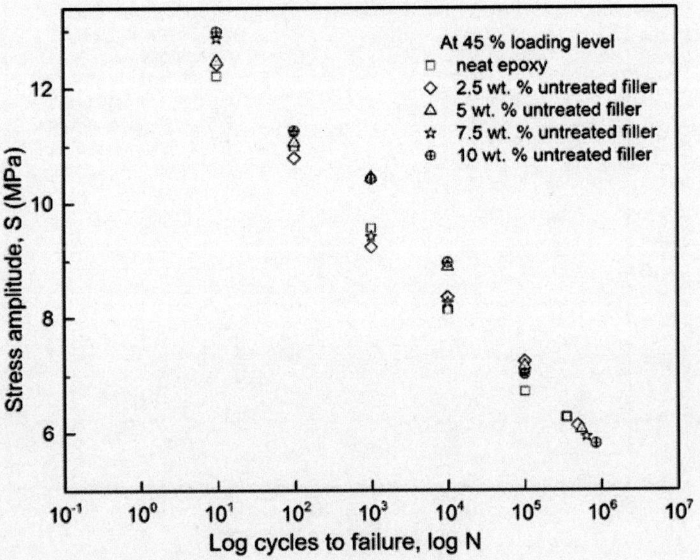

Fig. 7 *S–N* curve at 45% loading for untreated filler composite

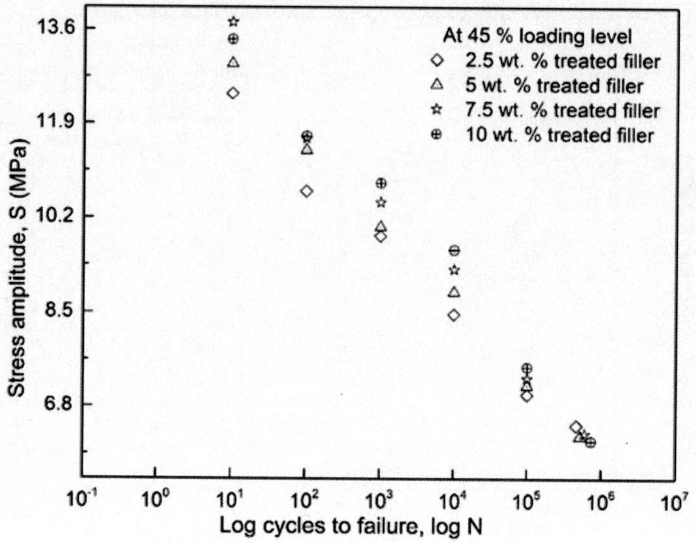

Fig. 8 *S–N* curve at 45% loading for treated filler composite

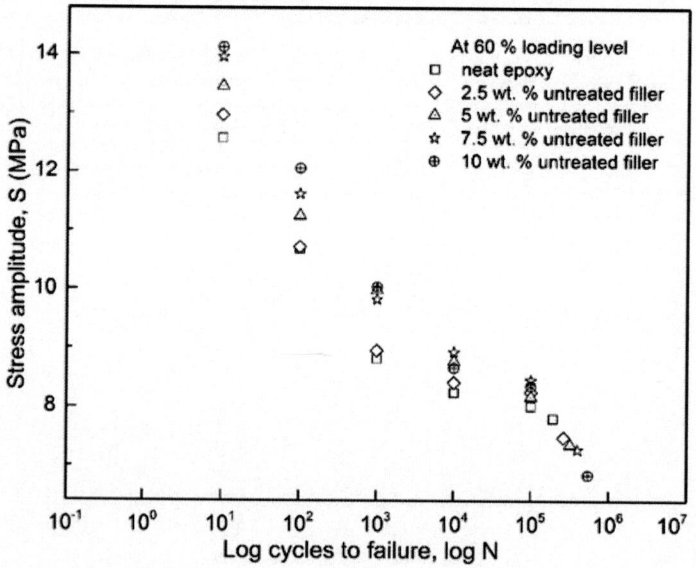

Fig. 9 *S–N* curve at 60% loading for untreated filler composite

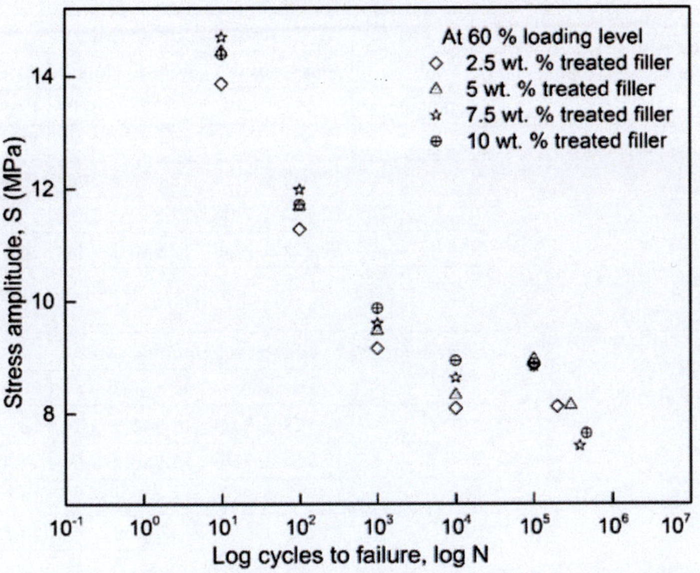

Fig. 10 *S–N* curve at 60% loading for treated filler composite

wt%, the fatigue strength continuously increases up to 7.5%, but the enhancement of stress from 5 wt% filler to 7.5 wt% is marginal. Nevertheless, after 7.5% of filler content, fatigue strength reduced for almost all the loading conditions. The decline in the strength is due to the agglomeration of fillers at higher filler content (10%) resulting in poor stress transfer. The filler dispersion in the epoxy matrix can be observed from scanning electron micrographs for 2.5 and 7.5 wt% of filler content in Fig. 11a, b, respectively. With an increase in filler content, the assemblage of fillers within the resin and hardener takes place. Also, the fillers at higher content in the matrix tend to form clusters and they engulf. The problem of agglomeration can be somewhat overcome by treating the fillers with aqueous NaOH solution. It acts as stress concentration resulting in debonding between the fillers and the matrix. The presence of voids is also a major factor that results in poor fatigue property. These imperfections act as stress concentration points subsequently shortening the cyclic stress values.

Tables 5 and 6 present the fatigue life in terms of the number of cycles to failure for untreated and treated filler composite samples, respectively. The neat epoxy specimen failed at 643,240, 346,860, and 194,280 number of cycles, respectively, for 30, 45, and 60% of loading level. At the same amount of loading levels, fatigue life was increased to 789,680 cycles, 486,780 cycles, and 264,740 cycles for 2.5% of untreated coir filler composite samples. Therefore, the filler addition in the epoxy matrix enhanced its fatigue life through 23–40% for different loading levels. The magnitude of failure cycles for 5 wt% filler samples have improved to 322,140 cycles, 543,380 cycles, and 831,600 cycles corresponding to 60, 45, and 30% of

Table 3 Cyclic stress mean values along with standard deviation for untreated filler composite

Filler content (wt%)	Cyclic stress amplitude (MPa)		
	30% loading	45% loading	60% loading
Neat epoxy	11.72 ± 2.49	12.221 ± 3.49	12.579 ± 2.94
2.5	11.991 ± 3.92	12.421 ± 3.81	12.967 ± 3.72
5	12.042 ± 3.28	12.484 ± 3.52	13.462 ± 3.07
7.5	12.104 ± 3.06	12.872 ± 3.2	13.962 ± 3.65
10	11.992 ± 3.38	12.986 ± 3.54	14.126 ± 3.66

Table 4 Cyclic stress mean values along with standard deviation for treated filler composite

Filler content (wt%)	Cyclic stress amplitude (MPa)		
	30% loading	45% loading	60% loading
2.5	11.928 ± 3.69	12.446 ± 3.75	13.864 ± 3.71
5	12.164 ± 3.76	12.986 ± 3.64	14.424 ± 3.37
7.5	12.926 ± 3.33	13.729 ± 3.27	14.687 ± 3.32
10	12.147 ± 2.7	13.426 ±	14.387 ± 3.68

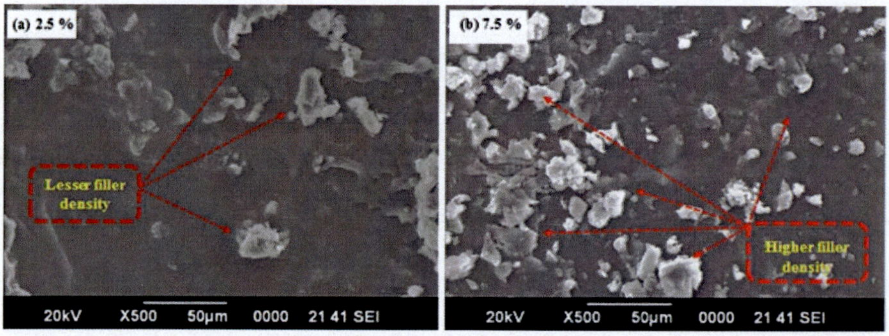

Fig. 11 Scanning electron micrographs of **a** 2.5 wt% and **b** 7.5 wt% filler composite samples

loading level. Corresponding to 60, 45, and 30% loading levels, the specimen failed at the end of 399,860 cycles, 632,340 cycles, and 942,380 cycles for 7.5 wt% of filler content specimens. Further addition of coir filler increased the fatigue life cycle to 542,420 cycles, 843,280 cycles, and 946,260 cycles with decreasing loading levels, respectively. The reason for the increase in fatigue life with coir filler addition might be attributed to the restrained crack initiation and propagation at higher filler content. The fatigue failure consists of the initiation of micro-cracks. This further results in uninterrupted propagation due to the presence of internal flaws in the material [28]. However, the filler particles in the matrix divert the crack propagation path, thus delaying the occurrence of material failure.

Table 5 Effect of loading level on fatigue life for untreated filler composite

Filler content (wt%)	Fatigue life (No. of cycles)		
	30% loading	45% loading	60% loading
Neat epoxy	643,240	346,860	194,280
2.5	789,680	486,780	264,740
5	831,600	543,380	322,140
7.5	942,380	632,240	399,860
10	946,260	843,280	542,420

Table 6 Effect of loading level on fatigue life for treated filler composite

Filler content (wt%)	Fatigue life (No. of cycles)		
	30% loading	45% loading	60% loading
2.5	778,120	462,640	198,480
5	814,820	516,480	294,640
7.5	906,480	598,440	384,240
10	Run out	728,840	468,420

3.6 *Effect of Chemical Treatment*

The influence of alkali treatment of coir fillers on the fatigue life of prepared composite materials is presented in Table 7. The observation can be built from the demonstrated experimental data that at 45% of the loading level; the alkali treatment brought a loss of 5–13% on the fatigue life of composite material. However, the chemical treatment of coir fillers resulted in the improvement of fatigue resistance to some extent as depicted in *S–N* curves. The shortened fatigue life could be due to the excessive fibrillation of treated coir fillers. It can be concluded that the chemical treatment of the fillers has improved the load bearing capacity and cyclic stress. The improvement of cyclic stress can be justified by the reduction of shear stress induced micro damage in the interface of matrix and filler as reported in previous research work [29]. Due to the removal of lignin and breakdown of hemicellulose, the fillers become more homogeneous by eliminating microvoids, and hence, the stress transfer capacity improves. The cellulose chains are also separated and get reoriented. This results in the release of initial strains of the fillers and subsequent re-adjustments to the chains. On the other hand, a considerable reduction in fatigue life for treated filler composite is observed at medium and high loading levels.

3.7 *Effect of Loading Levels*

The effect of loading levels on the fatigue life of composite samples can be observed from Tables 5 and 6. When the loading level was increased from 30 to 45%, a drop in

Table 7 Effect of chemical treatment on fatigue life at 45% loading condition

Filler content (wt%)	Fatigue life (No. of cycles)	
	Untreated filler composite	Treated filler composite
2.5	486,780	462,640
5	543,380	516,480
7.5	632,240	598,440
10	843,280	728,840

fatigue life was witnessed. This drop varied from 11 to 46% for 10 wt% of untreated filler content to neat epoxy samples, respectively. The further increase in loading level to 60%, resulted in the additional fall in fatigue life of the material. The reduced fatigue life at a higher loading level is associated with the allocation and reallocation of the molecules when subjected to a series of cycles. Similar findings were found in the works of other researchers [30, 31]. This also attributes to the factor that for lower number of cycles the variation of cyclic stresses is high. However, for higher number of cycles ($> 10^5$), there is no significant change in the values of cyclic stresses. The *S–N* curves can be divided into two stages—a zone with constant slope at the beginning of the curves, the other with gradual increasing slopes for a higher number of cycles. There is a steep increase in the slope of *S–N* curve at around 10^4 cycles. For low number of cycles, the rate of variation of cyclic stress is very high. As the number of cycles increases, the rate of fall in stress decreases. Matrix microvoids are formed at the initial stages of loading that gives rise to stiffness reduction [32]. It indicates that the samples are approaching their endurance limit. However, unlike metals, it is difficult to calculate the endurance limit of polymeric natural composite due to the complexity in the behavior of matrix and filler. Fatigue failure of filler-reinforced polymer composite materials seems to be a process of matrix crack initiation and growth, significant loss of creep and modulus. Other factors include accelerated filler debonding with the matrix and filler pull-out [16, 28]. It gives rise to crack initiation and propagation finally leading to failure of the composites samples.

4 Conclusion

This experimental investigation was conducted to study the effect of coir filler content, alkali treatment, and loading level on the fatigue life of coir–epoxy polymer composite. The salient conclusions of this work are outlined as follows:

- After 8 h of alkali treatment, the loss occurred in the weight of the filler is about 1.6% and is mainly due to the removal of lignin and other waxy particles from the filler surface. Aqueous NaOH alters the orientation and order of the highly packed crystalline cellulose, thereby forming an amorphous region.

- The presence of void content was continuously increased with an addition of coir fillers and its maximum value (2.94%) was observed for 10% of untreated filler content. The alkali treatment brought a drop in void contents in composite samples.
- The ultimate tensile strength (UTS) of the composite material was found to be increasing with coir filler content up to 7.5 and 5 wt% of untreated and treated filler amounts. The maximum value of UTS in the case of untreated and treated filler composites is 22 MPa and 25 MPa, respectively.
- It was observed that with the increase of coir filler content, both load bearing capacity and fatigue life of composite samples increase.
- The loading levels of 30% was observed to be optimum for achieving the highest fatigue life for 10 wt% of filler content. Moreover, the chemical treatment of the fillers improved the load bearing capacity but reduction in fatigue life was observed at all three loading levels.

This material can be used for domestic and industrial applications where the product is subjected to cyclic loading, i.e., environmental temperature fluctuations. The maximum value of UTS is 25 MPa; life cycle more than 10^6 cycles and cyclic stress 15 MPa, which signifies the impending applicability of the developed material.

Acknowledgements The authors acknowledge the facilities provided by Machine element laboratory, NIT Silchar and CIF-BIT Mesra, Jharkhand for conducting experiments. The authors also wish to acknowledge the support provided by Ministry of Education, India for supporting the research of the first and second author.

References

1. Pickering KL, Efendy MA, Le TM (2016) A review of recent developments in natural fibre composites and their mechanical performance. Compos Part A Appl Sci Manuf 83:98–112
2. Yashas Gowda TG, Sanjay MR, Subrahmanya Bhat K, Madhu P, Senthamaraikannan P, Yogesha B (2018) Polymer matrix-natural fiber composites: an overview. Cogent Eng 5(1):1446667
3. Arpitha GR, Sanjay MR, Senthamaraikannan P, Barile C, Yogesha B (2017) Hybridization effect of sisal/glass/epoxy/filler based woven fabric reinforced composites. Exp Tech 41(6):577–584
4. Sanjay MR, Madhu P, Jawaid M, Senthamaraikannan P, Senthil S, Pradeep S (2018) Characterization and properties of natural fiber polymer composites: a comprehensive review. J Clean Prod 172:566–581
5. Doan TTL, Gao SL, Mader E (2006) Jute/polypropylene composites I. Effect of matrix modification. Compos Sci Technol 66(7–8):952–963
6. Abdelmouleh M, Boufis S, Belgacem MN, Dufresne A (2007) Short natural-fibre reinforced polyethylene and natural rubber composites: effect of silane coupling agents and fibres loading. Compos Sci Technol 67(7–8):1627–1639
7. Okubo K, Fujii T, Thostenson ET (2009) Multi-scale hybrid biocomposite: processing and mechanical characterization of bamboo fiber reinforced PLA with microfibrillated cellulose. Compos Part A Appl Sci Manuf 40(4):469–475

8. Graupner N, Herrmann AS, Mussig J (2009) Natural and man-made cellulose fibre-reinforced poly(lactic acid) (PLA) composites: an overview about mechanical characteristics and application areas. Compos Part A Appl Sci Manuf 40(6–7):810–821
9. Venkateshwaran N, Perumal AE, Arunsundaranayagam D (2013) Fiber surface treatment and its effect on mechanical and visco-elastic behaviour of banana/epoxy composite. Mater Des 47:151–159
10. Komuraiah A, Kumar NS, Prasad BD (2014) Chemical composition of natural fibers and its influence on their mechanical properties. Mech Compos Mater 50(3):359–376
11. Fiore V, Di Bella G, Valenza A (2015) The effect of alkaline treatment on mechanical properties of kenaf fibers and their epoxy composites. Compos Part B Eng 68:14–21
12. Komal UK, Verma V, Ashwani T, Verma N, Singh I (2018) Effect of chemical treatment on thermal, mechanical and degradation behavior of banana fiber reinforced polymer composites. J Nat Fibers:1–13
13. Orue A, Jauregi A, Unsuain U, Labidi J, Eceiza A, Arbelaiz A (2016) The effect of alkaline and silane treatments on mechanical properties and breakage of sisal fibers and poly (lactic acid)/sisal fiber composites. Compos Part A App Sci Manuf 84:186–195
14. Mwaikambo LY, Tucker N, Clark AJ (2007) Mechanical properties of hemp-fibre-reinforced euphorbia composites. Macromol Mater Eng 292(9):993–1000
15. Ray D, Sarkar BK, Rana AK, Bose NR (2001) Effect of alkali treated jute fibres on composite properties. B Mater Sci 24(2):129–135
16. Gamstedt EK, Talreja R (1999) Fatigue damage mechanisms in unidirectional carbon-fibre-reinforced plastics. J Mater Sci 34(11):2535–2546
17. Bizeul M, Bouvet C, Barrau J, Cuenca R (2010) Influence of woven ply degradation on fatigue crack growth in thin notched composites under tensile loading. Int J Fatigue 32(1):60–65
18. Caprino G, Giorleo G (1999) Fatigue lifetime of glass fabric/epoxy composites. Compos Part A App Sci Manuf 30(3):299–304
19. Liang S, Gning PB, Guillaumat L (2012) A comparative study of fatigue behaviour of flax/epoxy and glass/epoxy composites. Compos Sci Technol 72(5):535–543
20. Asgarinia S, Viriyasuthee C, Phillips S, Dube M, Baets J, Van Vuure A, Lessard L (2015) Tension–tension fatigue behaviour of woven flax/epoxy composites. J Reinf Plas Compos 34(11):857–867
21. Gassan J (2002) A study of fibre and interface parameters affecting the fatigue behaviour of natural fibre composites. Compos Part A Appl Sci Manuf 33(3):369–374
22. Yang R (2003) Alkali deweighting of acid-modified multicomponent copolyester fibre. Ind J fiber text res 28(3):343–347
23. Zhu S, Wu Y, Yu Z, Chen Q, Wu G, Yu F, Wang C, Jin S (2006) Microwave-assisted alkali pre-treatment of wheat straw and its enzymatic hydrolysis. Biosyst Eng 94(3):437–442
24. Kumar R, Bhowmik S, Kumar K (2017) Establishment and effect of constraint on different mechanical properties of bamboo filler reinforced epoxy composite. Int Polym Proc 32(3):308–315
25. Ehrenstein GW, Wurmb R (1977) Verstarkte Thermoplaste—Theorie und Praxis. Die Angewandte Makromolekulare Chemie 60(61):157–214
26. Kumar R, Kumar K, Bhowmik S (2014) Optimization of mechanical properties of epoxy based wood dust reinforced green composite using Taguchi method. Proc Mater Sci 5:688–696
27. Kumar R, Kumar K, Sahoo P, Bhowmik S (2014) Study of mechanical properties of wood dust reinforced epoxy composite. Proc Mater Sci 6:551-556
28. Dao KC (1982) Fatigue failure mechanisms in polymer composites. Polym Compos 3(1):12–17
29. Towo AN, Ansell MP (2008) Fatigue of sisal fibre reinforced composites: constant-life diagrams and hysteresis loop capture. Compos Sci Technol 68(3–4):915–924
30. Towo AN, Ansell MP (2008) Fatigue evaluation and dynamic mechanical thermal analysis of sisal fibre–thermosetting resin composites. Compos Sci Technol 68(3–4):925–932

31. Abdullah AH, Alias SK, Jenal N, Abdan K, Ali A (2012) Fatigue behavior of kenaf fibre reinforced epoxy composites. Eng J 16(5):105–113
32. Nouri H, Meraghni F, Lory P (2009) Fatigue damage model for injection-molded short glass fibre reinforced thermoplastics. Int J Fatigue 31(5):934–942

Parallel Kinematics-Based Mechanism and Its Industrial Application in CNC Machine Tool Development

Mayur Singh, Priyanka Duarah, Sourabh Narnaware, and Shrikrishna Nandkishor Joshi

1 Introduction

A parallel kinematic manipulator is a closed-loop kinematic chain mechanism that consists of one or multiple closed kinematic loops. The end effector and fixed base are connected by two or more kinematic chains, which create each loop. In recent times, because of its high mechanical rigidity, high acceleration, and high workload, the adoption of these types of manipulators, in the industries, is growing very fast. Various practical applications of these manipulators are found significant in sectors like space, medical science, flight and automobile motion simulators, high-speed multi-axis machining and assembly, medical parallel robot, general pick-and-place operation, robots in food industries, etc. The PKM-based machine was firstly shown at the 1994 International Manufacturing Technology in Chicago by two American machine tool companies: Giddings & Lewis and Ingersoll [1]. Hexapods and Tripods are terms used to describe PKMs having six and three degrees of freedom (DOF), respectively [2, 3]. In 1965, D. Stewart proposed the six-DOF parallel manipulators, and hence, the six-DOF PKMs are also named as Stewart platform. However, this six-DOF platform was initially proposed by Gough in the year around 1956 to 1957 to test tire [4]. Therefore, this mechanism is also referred to as the Stewart–Gough platform. One of the pioneers in the subject of parallel robotics, Klaus Cappel, is an American engineer who created an octahedral hexapod manipulator as a motion simulator and patented it in 1967 [5].

This paper presents a literature review of the existing PKM industrial applications as well as a Monte Carlo-based approach to obtaining a well-conditioned workspace of a six-degree-of-freedom parallel manipulator. The region in three-dimensional Cartesian space that can be reached by a point on the moving platform can be thought

M. Singh · P. Duarah · S. Narnaware · S. N. Joshi (✉)
Department of Mechanical Engineering, Indian Institute of Technology Guwahati, Guwahati, Assam 781039, India
e-mail: snj@iitg.ac.in

S. N. Joshi et al. (eds.), *Low Cost Manufacturing Technologies*,
https://doi.org/10.1007/978-981-19-8452-5_18

of as the parallel robot's workspace, with the point of interest being the center of the moving platform. The main advantage of using the Monte Carlo method involves solving only the inverse kinematics problem for parallel manipulators.

2 Comparison of Serial Manipulator and Parallel Manipulator

Serial and parallel manipulators both have a wide range of industrial applications due to their unique characteristics. Serial Kinematic Manipulators offer a large dexterous workspace, and hence, they are widely adopted in the industries [6]. But, due to their cantilever kinematic structure, they are prone to bending at high load and they vibrate at high speed. This leads to a lack of precision and lesser load capacity. In parallel controllers, since the load is shared with all parallel links, there is a high load capacity. Also, these types of manipulators show high structural stiffness, high precision, high acceleration, and low inertia. In a parallel kinematic, multi-axis system, all actuator acts without delay on one moving platform. However, link interferences and limitations brought on by universal and spherical joints reduce the workspace's dexterity. On the other hand, the use of multi-DOF universal and spherical joints in PKM avoids the adverse effect of bending and torsion forces by limiting the load into compressive and tensile only [5]. Unlike serial manipulator, in case of PKM, presence of passive joint also can be found. Ultimately, this makes the kinematic analysis of PKM far different than serial ones. The inverse kinematic analysis of a PKM is quite simple and straightforward, and forward kinematic analysis is very much complex, while, in case of serial manipulator, it has been seen opposite. Another great advantage of PKM is that its error does not get accumulated rather it gets averaged. And hence, PKM-based robots are found to be more accurate than serial manipulator-based robots.

3 History of Industrial Applications of PKM

It has been observed that the majority of robotic arms used in industrial manufacturing utilize serial technology. But in the last few decades, parallel manipulators have been explored and developed extensively, both theoretically and practically, in academics as well as in industries. After reviewing a number of academic papers, it appears that Gwinnett, who received a patent in 1931 for a motion platform based on a spherical parallel, created the first real-world use for a parallel manipulator. The motion platform developed by Gwinnett was designed for use in the entertainment business [7].

Fig. 1 Universal tire testing machine of Gough [https://www.parallemic.org/Reviews/Review007.html (as seen on 17.5.2022)]

Willard L.V. Pollard received a patent in 1942 for a creative industrial parallel robot design that is regarded as the first PKM-based spatial industrial robot [6]. The design was intended to build a spray painting robot, but it was never built.

In 1954, Dr. Eric Gough who was known to be an employee of the Dunlop Rubber Company in England had designed and developed a six-degree-of-freedom octahedral hexapod for the universal tire testing machine for the first time [4]. As depicted in Fig. 1, this parallel manipulator was employed as a universal tire testing machine.

Later, Mr. Stewart had proposed the use of such a framework for flight simulators through his paper, which was published in the year 1965 [7]. The same architecture was concurrently proposed by Klaus Cappel as a motion simulator mechanism [7]. The Clavel-proposed Delta robot is one of the most successful parallel robot designs, besides the Gough platform [8].

Karl-Erik Neumann invented and manufactured a new type of robot in 1987 called the parallel kinematic robot (PKR) [9]. This kind of robot has three or more parallel linear axes that move in the same direction. It has three prismatic actuators: two of which regulate the rotational axis and the third controls the translational degree of freedom of the mobile platform.

Applications of parallel manipulators are quite promising in the field of medical science also. There are many applications in medical fields that involve robotics technology, such as injecting electrodes, drilling bone, medical transportation, and performing various other tests. Wall painting, shipbuilding, bridge construction,

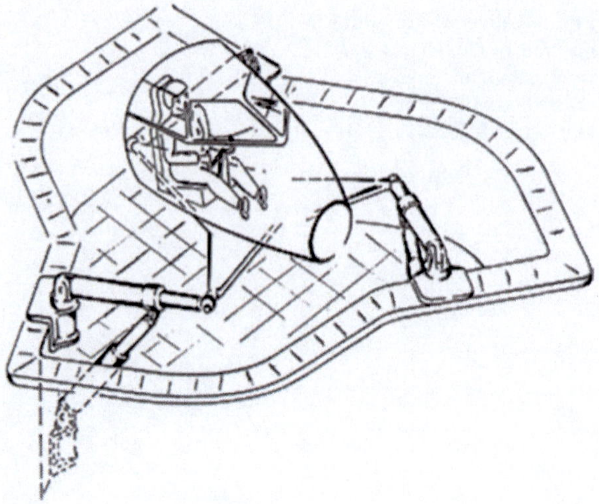

Fig. 2 Original Stewart platform proposed as a flight simulator [https://www.parallemic.org/Reviews/Review007.html (as seen on 17.5.2022)]

aircraft maintenance, ship-to-ship freight handling, and other applications require service robots that can move in a vertical plane, and hence, PKM can be a potential technology for these applications. Due to developments in computer technology and the development of advanced control systems, it can be said that parallel manipulators have recently become more practical and efficient technology for various high-demand industrial products like CNC-based multi-axis manufacturing machines, robots used in automobile industries, etc.

4 Computer Numerical Control (CNC)

Computer numerical control (CNC) machines arose from the advancement of numerical control (NC) machines. The machine control unit (MCU) understands and processes numerical control (NC), which is a series of coded instructions consisting of alphanumeric letters and symbols. To carry out machining operations on a workpiece, the machine's motors and controllers convert these commands into electrical current pulses. The machine tool's numbers, letters, and symbols serve as coded instructions that refer to certain lengths, positions, functions, or motions that the machine tool may decode as it processes the workpiece. NC and CNC machines are currently widely employed in a wide range of sectors, from small to large [10].

Modern CNC systems use computer-aided design (CAD) and computer-aided manufacturing (CAM) software to automate the entire component design process. These applications generate a computer file that is processed by a post-processor to extract the commands needed to run a specific machine and then fed into CNC machines for manufacturing. These machines perform numerous processes such as milling, drilling, sawing, and other operations with several tools in a single cell, saving time and increasing output rate and efficiency.

4.1 Stewart Platform and CNC Machine

The Stewart–Gough platform, often known as the hexapod, is one of the most well-known six-degree-of-freedom parallel kinematics manipulators. The mechanism is made up of a fixed base plate and a moving platform which are linked by six prismatic actuators via spherical and universal joints. The six actuators can be either linear or angular, and they connect the robot's bottom, which has no mobility, with the component that does the end effector [1]. Because of the closed-loop architecture, not all of the joints can be independently actuated. Inverse kinematics analysis is used to determine the lengths of the connections in terms of position, orientation, translation, and angular velocities. When compared to solving the forward kinematics equations, the inverse kinematics equation is much simpler to solve. By changing the link lengths, the orientation and positions of the moving plate can be controlled. The Stewart platform can be used to construct a CNC hexapod machine tool. A hexapod CNC machine can be an excellent machining tool that addresses various demands of the modern machining processes and can be modified according to the needs. Nowadays, CNC machines are widely employed in several sectors like 3D printing, graphics printing on tee shirt, building construction, food sorting, etc. End-to-end component design is highly automated in modern CNC systems, using computer-aided design (CAD) and computer-aided manufacturing (CAM) programs.

4.2 Advantages of Stewart Platform-Based CNC

The Stewart platform is gaining favor in the design of multi-degree freedom machine tools due to its accurate movements. Machines with high speed, accuracy, rigidity, and multi-axial capabilities, as well as high-quality produced parts, are required for today's new generation. CNC hexapod machining has a lot of potential for transforming manufacturing processes. As listed below, there are various advantages to such a machine.

Six Degrees of Freedom

The hexapod is made up of two platforms joined by six struts that move back and forth between the mobile and fixed platforms. The coordinated motion of these six struts enables the spindle, which is mounted on the movable platform, to move in any direction. In addition to the traditional motion in orthogonal axis movements x, y, z (lateral, longitudinal, and vertical), this device can also move in the rotary α, β, γ (pitch, roll, and yaw) coordinates. This allows the spindle to reach unusual angles and geometrical features and provides a significant advantage over conventional CNC machining.

High Precision and Accuracy

When compared to traditional multi-axis positioning tools, the hexapod requires all six struts to change their lengths even if only one axis needs to be changed. If only one strut changes its length, all six coordinates (x, y, z) need to be adjusted. This common motion of movable platforms, which is dependent on all struts, ensures great precision and accuracy.

High Stiffness

A hexapod's high stiffness and rigidity of its components and all moving parts, such as joints and drive screws, are another important feature. It produces extremely high natural frequencies (500 Hz @ 10 kg load), which enables for extremely high-speed cutting operations and other machining goals. Furthermore, all of a hexapod's struts are solely subjected to forces in the longitudinal direction, preventing any bending effects in the struts and providing a significant advantage.

High Load to Weight Ratio

The hexapod's high nominal load to weight ratio is a key benefit. The weight of a load on the platform is spread about evenly among the six parallel links, implying that each link bears just one-sixth of the entire weight. The hexapod's struts function longitudinally, and the load applied to them exerts either tension or compression forces.

5 Workspace Analysis By Monte Carlo Method

The Monte Carlo approach uses random sampling to solve mathematical problems numerically. Any phenomenon that is influenced by random variables can be simulated using the Monte Carlo approach [11]. The primary disadvantage of parallel manipulators is their limited workspace. In serial manipulators, by using the geometry of the manipulator, its Denavit–Hartenberg parameters and the actuated joint limits, the workspace could be easily determined. But in parallel or hybrid manipulators, in addition to the actuated joint values, the values of the passive joints are determined by solving the forward kinematics. The real solutions for the forward kinematics may not exist which in turn prohibits the parallel manipulator to assemble for given joint variables [12]. Additionally, the self-collision of the links and singularities may split the workspace and increase the complexity of determining the workspace. The approach is search based—a region in space is discretized and the discrete points are generated, the inverse kinematics is solved at generated discretized points, and then the joint variables are checked for violation of limits. The best result is obtained by discretizing the points finer. To get a well-conditioned workspace of a six-degree-of-freedom parallel manipulator, Monte Carlo-based technique is applied [12].

5.1 Monte Carlo Method

The integrals of any arbitrary functions (vector or scalar functions of smooth or non-smooth type) over any arbitrary domain could be evaluated using the Monte Carlo approach [12]. The integral

$$I = \int_{[0,1]^d} f(x)\mathrm{d}x \tag{1}$$

where $f(.)$ is a bounded real-valued function, can be obtained as $E(f(U))$, where $E(.)$ is the expectation of a variable taking a particular probabilistic value and $U = [u_1, u_2, \ldots, u_d]^T$ a $1 * d$ vector taking random values of $u_i \in [0, 1]\ \forall i = 1, 2, \ldots, d$ (Fig. 3).

The Monte Carlo method has been used to obtain the well-conditioned and reachable workspace of a closed-loop mechanism or parallel manipulator, by recognizing that the volume (or area) of a manipulator is an integration problem in $R^3(R^2)$. It can be implemented by formulating a function $f(j) = \begin{cases} 0 \\ 1 \end{cases}$, where $j = \{\theta_i, \phi_j\}^T$, $\forall i = 1, 2, \ldots, n$ actuated joint variables and $\forall j = 1, 2, \ldots, m$ passive joint variables and $m + n = d$.

The function f takes 0 or 1 depending on whether the given position and orientation of the end effector of manipulator are well conditioned, and the condition for inverse kinematic of the manipulator is satisfied at that position and orientation with all the joint values within permissible joint limits. The main advantage of using Monte Carlo method involves solving only the inverse kinematics problem for parallel manipulator, and the other conditions could be incorporated to ensure that the workspace is well conditioned and does not violate any restrictions.

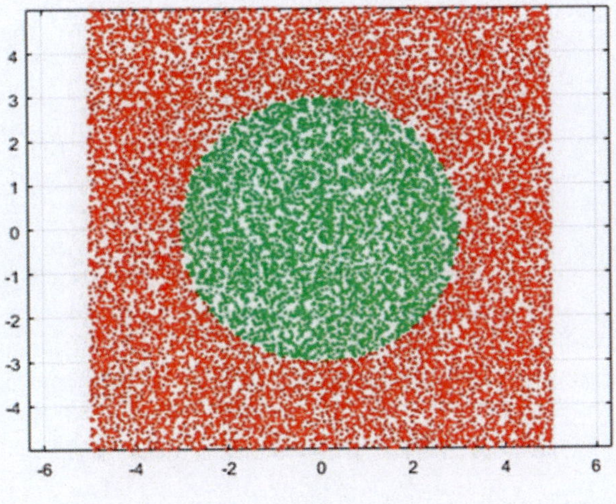

Fig. 3 Visualization of Monte Carlo search

5.2 Solution of Inverse Kinematics of Stewart Platform

A Stewart platform is shown in Fig. 4, where P denotes the position of the spherical joint on the top plate and B denotes the position of universal joint on the base plate. Let l denote the actuator length.

A leg of a Stewart platform is shown in Fig. 5, where a vector equation in this leg is applied in order to get the kinematics of a loop of a Stewart platform.

Rotation matrix can be written as

$$R_x = \begin{bmatrix} 1 & 0 & 0 \\ 0 & \cos\varphi_1 & -\sin\varphi_1 \\ 0 & \sin\varphi_1 & \cos\varphi_1 \end{bmatrix}$$

$$R_y = \begin{bmatrix} \cos\phi_1 & 0 & \sin\phi_1 \\ 0 & 1 & 0 \\ -\sin\phi_1 & 0 & \cos\phi_1 \end{bmatrix}$$

$$R_z = \begin{bmatrix} \cos\theta & -\sin\theta & 0 \\ \sin\theta & \cos\theta & 0 \\ 0 & 0 & 1 \end{bmatrix}$$

Let

^{B_0}t denote the location of the origin $\{P_0\}$ with respect to $\{B_0\}$,

$^{B_0}_{P_0}[R]$ Orientation of top platform with respect to fixed base,

Ψ Euler rotation about X-axis,

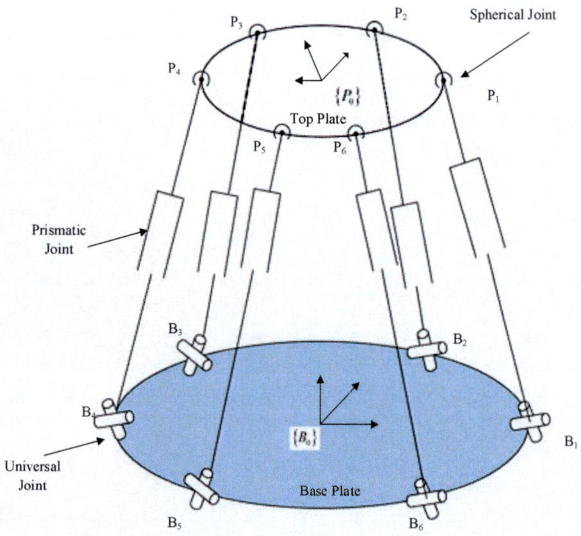

Fig. 4 Stewart platform

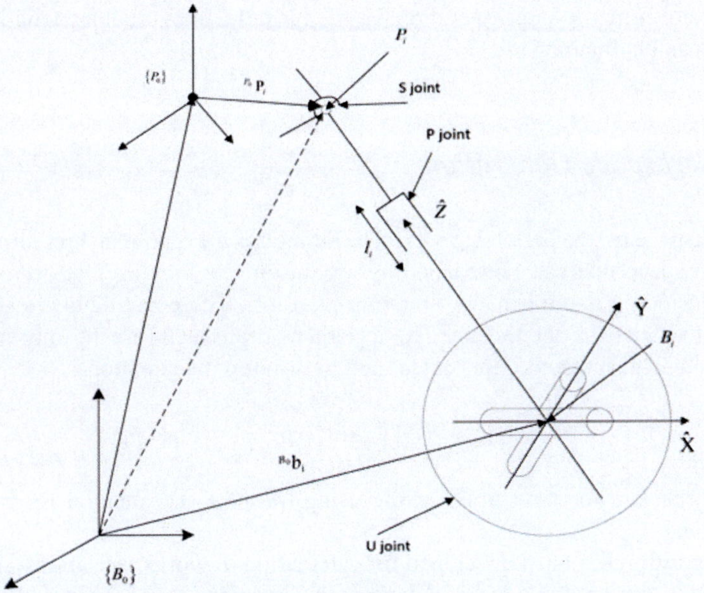

Fig. 5 Schematic diagram of a leg of Stewart platform

ϕ Euler rotation about Y-axis,
θ Euler rotation about Z-axis.

From Fig. 5, an arbitrary platform point P_i can be written in $\{B_0\}$ as

$$^{B_0}P_i = {}^{B_0}_{P_0}[R]^{P_0}P_i + {}^{B_0}t \tag{2}$$

The $^{P_0}P_i$ is a known constant vector in $\{P_0\}$.
The location of the base connection points $^{B_0}b_i$ is known.
From the known $^{B_0}_{P_0}[R]$ and translation vector ^{B_0}t, $^{B_0}P_i$ is obtained.

$$\left[R\left(\hat{Z}, \gamma_i\right)\right]^T \left[(x, y, z)^T - {}^{B_0}b_i\right] = \left[R\left(\hat{Y}, \phi_i\right)\right]\left[R\left(\hat{X}, \varphi_i\right)\right](0, 0, l_i)^T \tag{3}$$

$$\left[R\left(\hat{Z}, \gamma_i\right)\right]^T \left[(x, y, z)^T - {}^{B_0}b_1\right] = l_1 \begin{Bmatrix} \sin\phi_1 \cos\varphi_1 \\ -\sin\varphi_1 \\ \cos\phi_1 \sin\varphi_1 \end{Bmatrix} \tag{4}$$

After solving the above Eq. (4), the leg length for an actuator can be obtained as follows:

$$l_1 = \pm\sqrt{\left[(x, y, z)^T - {}^{B_0}b_1\right]^2} \tag{5}$$

By performing the same operation for each leg, the individual leg length for each actuator can be obtained.

5.3 Workspace Generation

The workspace for the parallel robot can be defined as the region in three-dimensional Cartesian space that can be reached by a point on the moving platform, the point being considered is the center of moving platform. The constraints are simply the link lengths that have been selected for a given workpiece and are the only constraints considered. The actuated joints calculated by solving the equation

$$\left[(x, y, z)^T - {}^{B_0}b_1\right]^2 = l_1^2\left(\sin^2\varphi_1 + \cos^2\varphi_1\right) \tag{6}$$

are then checked for joint limits while using Monte Carlo method for workspace generation.

An algorithm has been developed for calculation of workspace and work volume for various configurations of PKM-based CNC machine tool by using Monte Carlo method.

Following are the results obtained by simulating different models of Stewart platform. The random number was generated, and a number of random points taken were in order of 10^3 to 10^6. Following results are for 10^5 sampling points. The orientation of the top plate is considered to be horizontal.

Configuration 1

In this configuration, the diameter of base plate is 600 mm and size of top plate is 470 mm. The actuator joints are placed at 10 mm from the edge of the plate as shown in Fig. 6. On base plate, the actuators are located at 60° and are evenly spaced. The first actuator is at 30° with the horizontal reference axis as shown in Fig. 6 followed by the second actuator at 60° to the first and so on the next actuators are placed. While using Monte Carlo method, the joint constraint equation has been applied as the link length should fall in between 324 and 625 mm as these are the limits for our actuators.

On the top plate, the spherical joint is located at 60° and is evenly spaced. The first spherical joint is at 30° with the reference and followed by the second joint which is at 60° to the first and the third as shown in Fig. 7.

Assembly is done by connecting the top and base plate with the help of linear actuator. The connections are symmetrical, the first U-joint is connected to the first S-joint and so on as shown in Fig. 8.

Figure 9 shows the workspace for spatial Stewart platform. The orientation of the top plate is considered to be horizontal. Triangulation of the spatial plot of Fig. 10 is shown in Fig. 9.

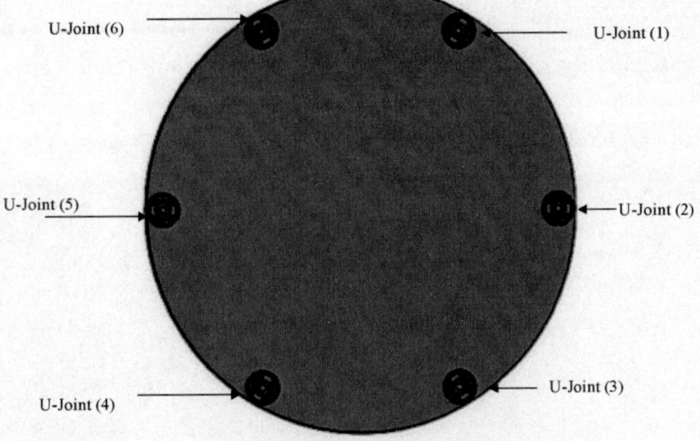

Fig. 6 Base plate for configuration 1

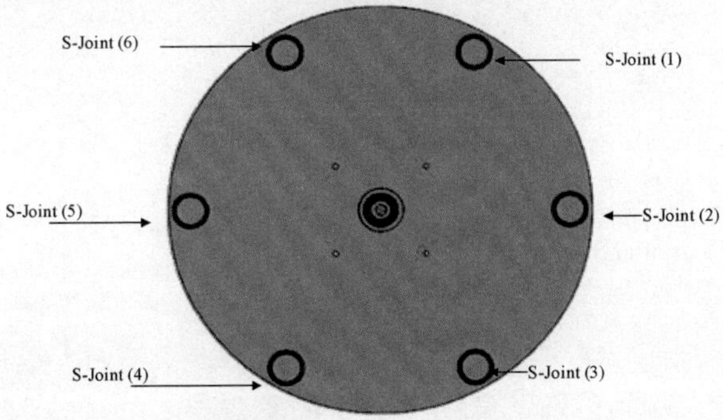

Fig. 7 Top plate for configuration 1

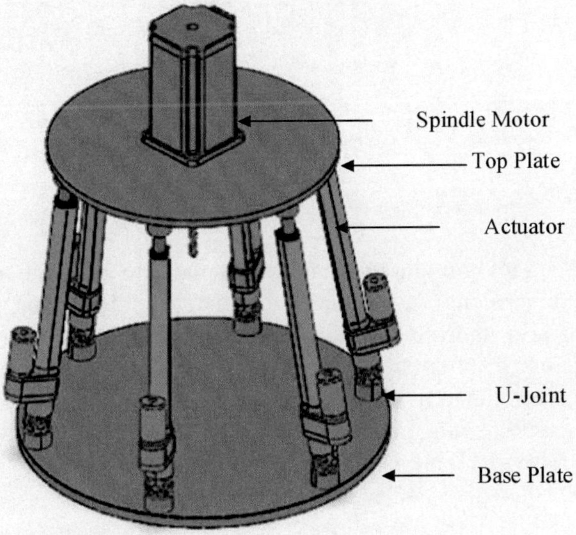

Fig. 8 Assembly for configuration 1 (isometric view)

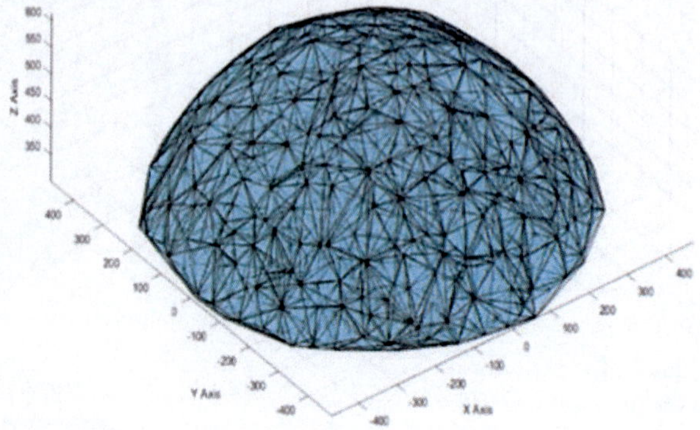

Fig. 9 Available workspace for configuration 1 (with triangulation)

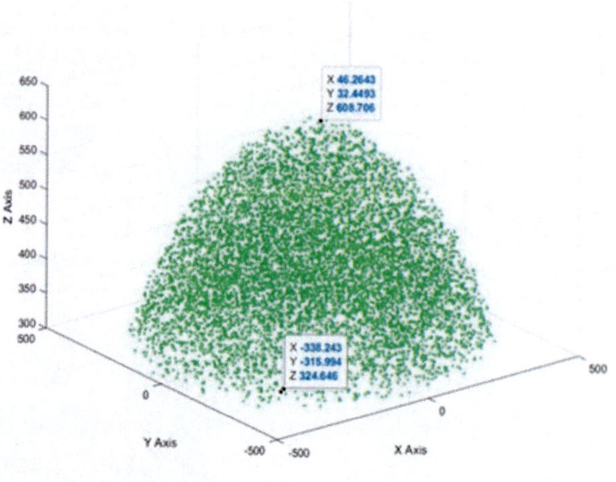

Fig. 10 Available workspace for configuration 1 (spatial Stewart platform)

This configuration of Stewart platform gives the maximum volume for the given dimensions of top and base plate and for given actuator stroke. Since Monte Carlo algorithm is a random number generation approach, we have done maximum number of simulation, but the variation in the workspace volume was found to be approximately same.

Following are the results obtained by simulation of MATLAB code for configuration 1 (Table 1).

Configuration 2

In this configuration, the diameter of base plate is 600 mm and size of top plate is 470 mm. The actuator joints are placed at 10 mm from the edge of the plate as shown in Fig. 11. On base plate, the actuators are located at 60° and are evenly spaced. The

Table 1 Simulation results for configuration 1

Sample points satisfying constraints	Volume of workspace in mm^3
8876	124,260,000
8949	125,286,000
8830	123,620,000
8908	124,710,000
8932	125,048,000
8856	123,984,000
8851	123,910,000
8862	124,068,000
8843	123,802,000
8962	125,468,000

first actuator is at 30° with the horizontal reference axis as shown in Fig. 11 followed by the second actuator at 60° to the first and so on the next actuators are placed. While using Monte Carlo method, the joint constraint equation has been applied as the link length should fall in between 324 and 625 mm as these are the limits for our actuators.

On the top plate, the spherical joint is located at 40° and 80° apart and is not evenly spaced. The first spherical joint is at 20° with the reference and followed by the second joint which is at 80° to the first and 40° to the third as shown in Fig. 12.

Assembly is done by connecting the top and base plate with the help of linear actuator. The connections are not symmetrical in this case; here the first U-joint is connected to the first S-joint which are located at 30° and 20° with the reference axis which is Y-axis on X–Y plane. This can be seen in Fig. 13.

Figure 14 shows the workspace for spatial Stewart platform. The orientation of the top plate is considered to be horizontal. Triangulation of the spatial plot of Fig. 15 is shown in Fig. 14.

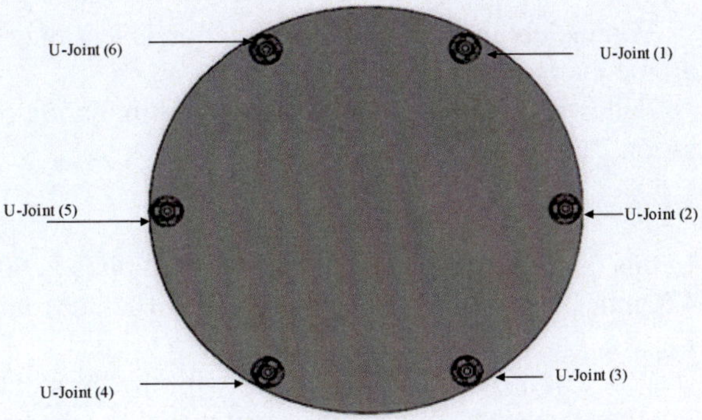

Fig. 11 Base plate for configuration 2

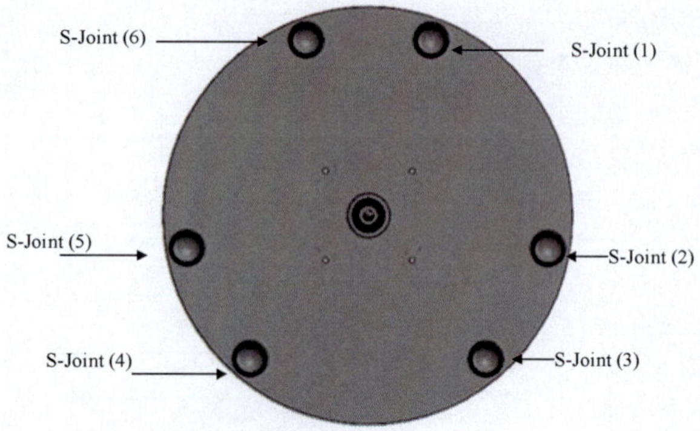

Fig. 12 Top plate for configuration 2

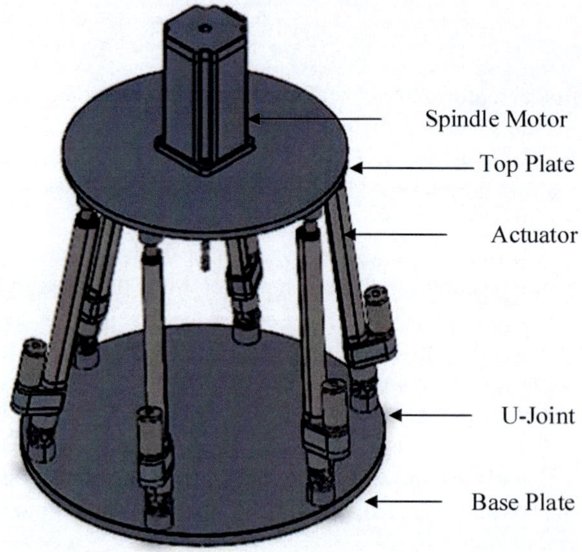

Fig. 13 Assembly for configuration 2 (isometric view)

Volume obtained in this configuration is less than that of configuration 1, where all the joints were equally spaced.

Following are the results obtained by simulation of MATLAB code for configuration 2 (Table 2).

Configuration 3

In this configuration, the diameter of base plate is 600 mm and size of top plate is 470 mm. The actuator joints are placed at 10 mm from the edge of the plate as shown Fig. 16. On base plate, the actuators are located at 40° and 80° and are not evenly spaced. The first actuator is at 40° with the horizontal reference axis as shown in Fig. 16 followed by the second actuator at 40° to the first and 80° to the third universal joint or actuator.

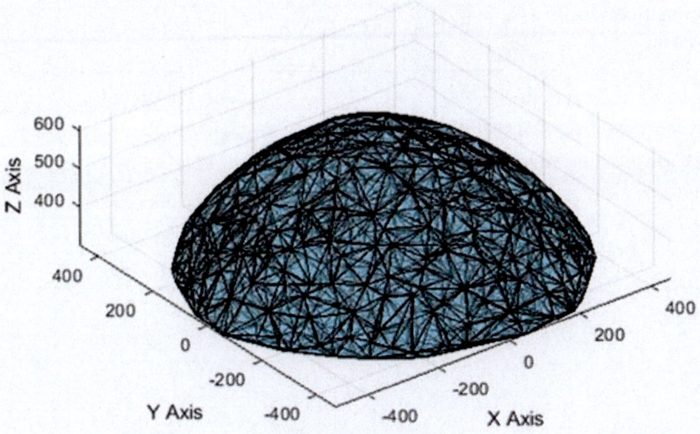

Fig. 14 Triangulation on the plot for configuration 2

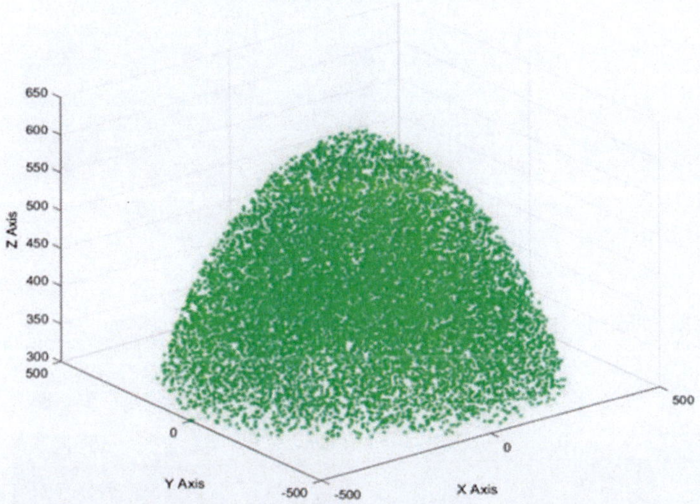

Fig. 15 Available workspace for configuration 2 (spatial Stewart platform)

On the top plate, the spherical joint is located at 40° and 80° apart and is not evenly spaced. The first spherical joint is at 20° with the reference and followed by the second joint which is at 80° to the first and 40° to the third as shown in Fig. 17.

Assembly is done by connecting the top and base plate with the help of linear actuator. The connections are not symmetrical in this case; here the first U-joint is connected to the first S-joint which are located at 20° and 40° with the reference axis which is *Y*-axis on *X*–*Y* plane. This can be seen in Fig. 18.

Table 2 Simulation results for configuration 2

Sample points satisfying constraints	Volume of workspace in mm^3
8534	119,476,000
8645	121,030,000
8652	121,128,000
8738	122,332,000
8683	121,562,000
8676	121,464,000
8615	120,610,000
8788	123,032,000

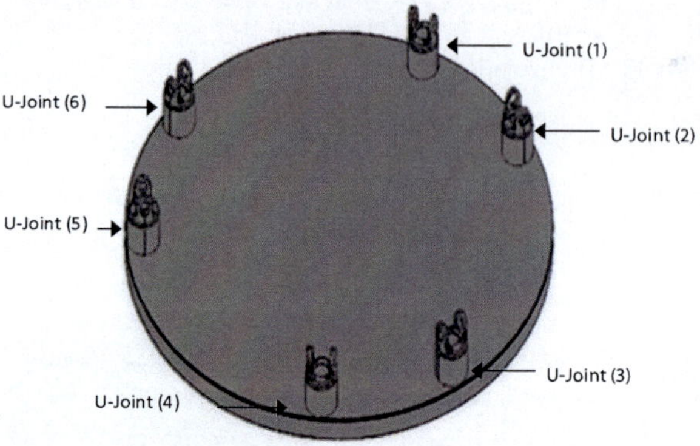

Fig. 16 Base plate for configuration 3 (isometric view)

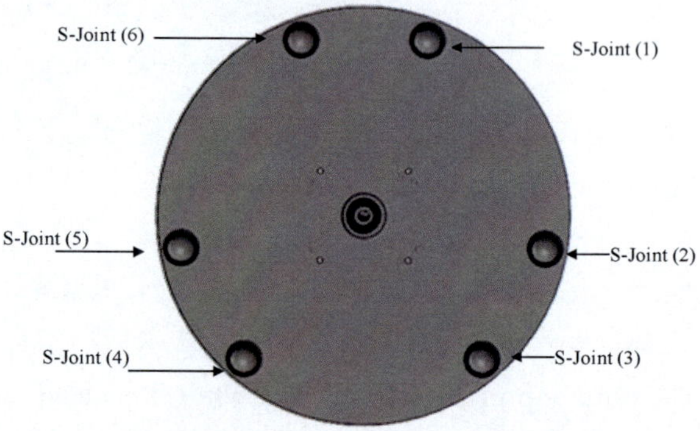

Fig. 17 Top plate for configuration 3

Figure 19 shows the workspace for spatial Stewart platform. The orientation of the top plate is considered to be horizontal. Triangulation of the spatial plot of Fig. 20 is shown in Fig. 19.

The volume for configuration 3 is least of all the three configurations.

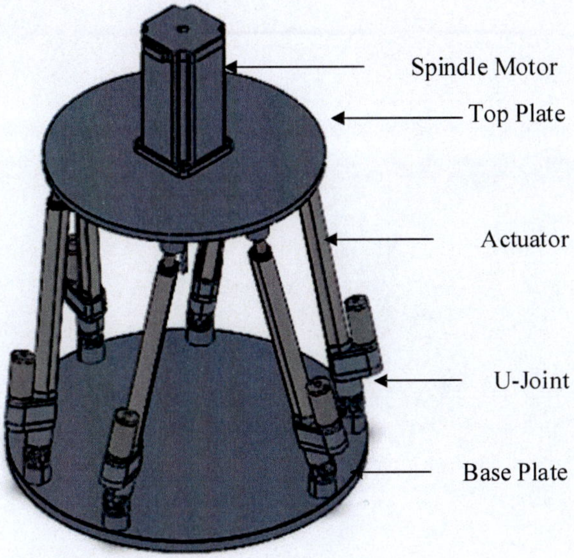

Fig. 18 Assembly for configuration 3 (isometric view)

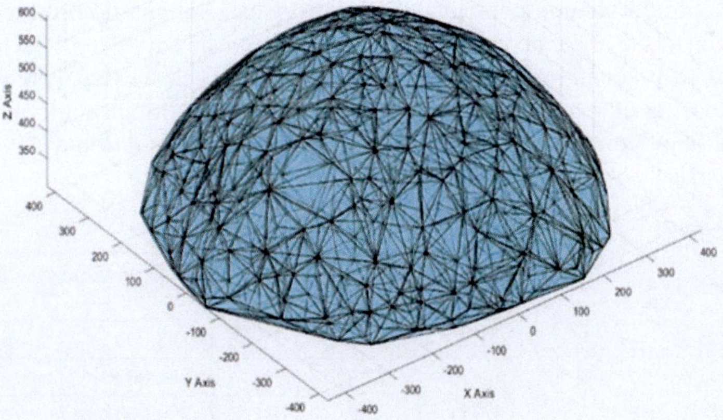

Fig. 19 Triangulation of the workspace for configuration 3

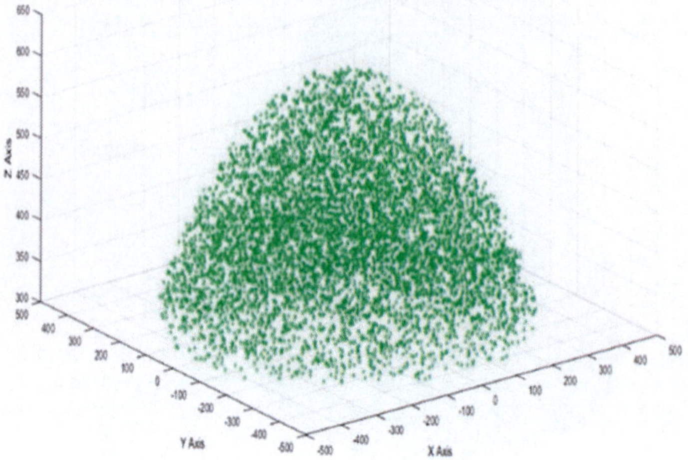

Fig. 20 Available workspace for configuration 3 (spatial Stewart platform)

Following are the results obtained by simulation of MATLAB code for configuration 3 (Table 3).

The volume obtained by configuration 1 is maximum, but due to its instability, configuration 2 will be selected for the PKM-based CNC machine tool. Depending on the maximum volume computed for a particular configuration, the orientation and position of the linear actuator on top and bottom plate have been selected, which comes out to be configuration 2, i.e., 60° equally spaced universal joint and 40° and 80° spacing of spherical joint on top plate. The first S-joint should be placed at 20° from the reference axis, and the first U-joint should be placed at 30° from the reference axis.

Table 3 Simulation results for configuration 3

Sample points satisfying constraints	Volume of workspace in mm^3
6942	97,188,000
6883	96,362,000
6771	94,794,000
6892	96,488,000
6957	97,398,000
6910	96,740,000
6757	94,598,000
6947	97,258,000

6 Conclusion

This paper presents Parallel Kinematics Manipulator and its advantages over Serial Kinematic Manipulator. A literature review on the existing industrial applications of PKM has been done. A detailed discussion about the computer numerical control (CNC) and Stewart platform-based CNC machines has also been brought off in this research paper. The advantages of Stewart platform have also discussed. Moreover, three different configurations of Stewart platform have been compared and simulated using MATLAB coding to calculate the workspace volume and position and orientation of the linear actuators, thereby selecting the most feasible configuration required for our suggested tabletop CNC machine configuration. Monte Carlo-based method to compute the workspace using inverse kinematics analysis and work volume for different configurations has been carried out in this paper. It is effective to solve for the conditioned workspace area and manipulator volumes using the Monte Carlo approach. The Monte Carlo-based algorithm and computed volumes for three-dimensional workspaces will aid to the design and fabrication of the suggested simple and low-cost tabletop CNC machine configuration.

References

1. Stan SD, Maties V, Balan R (2008) Optimal design of parallel kinematics machines with 2 degrees of freedom. In Parallel Manipulators, Towards New Applications. IntechOpen
2. Herve JM, Sparacino F (1991) Structural synthesis of 'parallel' robots generating spatial translation. Fifth international conference on advanced robotics robots in unstructured environments, Pisa, Italy, vol. 1, 808–813. https://doi.org/10.1109/ICAR.1991.240575
3. Kong X, Gosselin CM (2002) Type synthesis of linear translational parallel manipulators. Advances in robot kinematics—Theory and applications, Lenarcˇicˇ J. and Thomas F. Eds., Kluwer Academic Publishers, pp. 411–420
4. Stewart D (1965) A platform with six degrees of freedom. Proc Inst Mech Eng 180(1):371–386
5. Patel YD, George PM (2012) Parallel manipulators applications—a survey. Mod Mech Eng 02(03):57–64
6. Alazard D, Chretien JP (1994) Dexterous manipulation in space: Comparison between serial and parallel concepts, IFAC Proceedings Volumes, 27(13): 411–418
7. Gallardo-Alvarado J (2016) Kinematic analysis of parallel manipulators by algebraic screw theory, Springer Cham, pp 1–377. https://doi.org/10.1007/978-3-319-31126-5
8. Staicu S, Carp-Ciocardia DC (2003) Dynamic analysis of Clavel's delta parallel robot. Proc IEEE Int Conf Robot Autom 3(May):4116–4121
9. Angeles J, Morozov A, Bai S (2005) A novel parallel-kinematics machine tool, Proceedings of The fifth international workshop on advanced manufacturing technologies AMT2005 At: London, Ontario, Canada 1–6
10. Rolland L (2015) Path planning kinematics simulation of CNC machine tools based on parallel manipulators. Mech Mach Sci 29:147–192
11. Rastegar J (1990) Manipulation workspace analysis using the Monte Carlo Method. Mech Mach Theory 25(2):233–239
12. Chaudhury AN, Ghosal A (2017) Optimum design of multi-degree-of-freedom closed-loop mechanisms and parallel manipulators for a prescribed workspace using Monte Carlo method. Mech Mach Theory 118:115–138